PRINCIPLES AND APPLICATIONS OF
Palaeomagnetism

PRINCIPLES AND APPLICATIONS OF

Palaeomagnetism

D. H. TARLING

Department of Geophysics and Planetary Physics,
The University of Newcastle upon Tyne

CHAPMAN AND HALL
11 NEW FETTER LANE, LONDON EC4

First published 1971
© 1971 D. H. Tarling
Set by Santype Ltd.
(Coldtype Division)
Printed in Great Britain by
Cox & Wyman Ltd.
Fakenham, Norfolk

SBN 412 10910 7

DHG
T

322

Distributed in the U.S.A. by
Barnes & Noble Inc.

Contents

Preface

1   Introduction   ix

1.1   *Introduction*   1
1.2   *The early history of palaeomagnetism and geomagnetism*   1

2   The physical basis

2.1   *Introduction*   5
2.2   *Magnetism on an atomic scale*   5
2.3   *Domains and macroscopic magnetisation*   9
2.4   *The effect of time, temperature and volume changes*   14
2.5   *Pressure effects and magnetic anisotropy*   18

3   The commoner magnetic minerals and their identification

3.1   *Introduction*   21
3.2   *The iron-titanium oxides in igneous and metamorphic rocks*   21
3.3   *The magnetic minerals in sedimentary rocks*   27
3.4   *Less common magnetic minerals*   28
3.5   *The identification of magnetic minerals*   29

4   The magnetisation of rocks and its physical analysis

4.1   *Introduction*   32
4.2   *Primary magnetisation*   33
4.3   *Metamorphic rocks*   37
4.4   *Secondary magnetisation*   38
4.5   *The stability of remanence*   40

4.6	*The age of remanence*	48
4.7	*Magnetic anisotropy and inhomogeneity*	52
4.8	*Summary and comments*	53

5 Sampling and measurements

5.1	*Introduction*	55
5.2	*Sampling rock formations*	55
5.3	*Orientation of samples*	57
5.4	*Measurement of remanence, susceptibility, anisotropy and inhomogeneity*	60
5.5	*Demagnetisation*	66
5.6	*Field free space*	69
5.7	*The accuracy of palaeomagnetic techniques*	70

6 Statistical analyses

6.1	*Introduction*	
6.2	*Average directions of magnetisation and pole positions*	72
6.3	*Estimates of precision and scatter of palaeomagnetic directions*	72
		75
6.4	*Analysis of groups of directions*	83
6.5	*Levels of statistical analyses*	84
6.6	*Measurements of stability*	86
6.7	*Susceptibility and intensity of magnetisation*	87
6.8	*Discussion*	90

7 Geomagnetic applications ←

7.1	*Introduction*	92
7.2	*The present geomagnetic field*	93
7.3	*Secular variation*	98
7.4	*The intensity of the ancient field*	104
7.5	*The average nature of the ancient geomagnetic field*	107
7.6	*Palaeomagnetism and the Earth's interior*	112

8 **Reversals of magnetisation**

8.1	*Introduction*	114
8.2	*Self-reversal mechanisms*	114
8.3	*Correlations between petrology and polarity*	117
8.4	*Correlations between polarity and the age of rocks*	119
8.5	*Polarity transition zones*	121
8.6	*Geomagnetic reversals and the polarity time scale*	123
8.7	*Reversals and evolution rates*	127

9 **Geological applications**

9.1	*Introduction*	129
9.2	*The dating of rocks*	129
9.3	*Tectonic and structural applications*	136
9.4	*Palaeomagnetism and palaeolatitudes*	147
9.5	*The geological history of rocks*	150

Selected bibliography 152

Index 161

Preface

I have written this book as an introduction to palaeomagnetism for final year undergraduate students of geology or physics and for postgraduate students who are unfamiliar with this field of study, although certain statistical aspects are mainly of postgraduate interest. The subject incorporates three main disciplines; the study of the physical processes involved during the acquisition of magnetisation (rock magnetism), investigation of the nature and origin of the Earth's ancient magnetic field (geomagnetism) and the study of the age and tectonic history of rocks (geology-geophysics). I have attempted to give an understandable and balanced account of the relevant portions of these disciplines while amplifying certain aspects which are particularly relevant to palaeomagnetism. The book does not provide a comprehensive review of results but is intended to indicate the ways in which a variety of problems can be tackled using palaeomagnetic techniques.

The book could not have been written without the co-operation of many people, particularly past and present colleagues in the Department of Geophysics and Planetary Physics of the University of Newcastle upon Tyne and the Department of Geophysics and Geochemistry of the Australian National University. In particular, Drs W. O'Reilly and A. Stephenson assisted in the discussion of physical aspects and Drs E. A. Hailwood and J. H. Parry read and suggested corrections to the manuscript. Most of all I should like to thank my wife who made numerous contributions and improvements to my original text.

Newcastle upon Tyne   March 1971

UNITS

Standard SI units have been used throughout this book with the exception of magnetic units for which the cgs gauss and oersted units have been retained. This has been done because there is, as yet, no international agreement as to whether magnetic moment in emu should be interpreted as ampere turns square metre or webers metre. The SI conversions are given below:

magnetic field intensity (H)	1 oersted	$= 10^3/4\pi$ A m^{-1}
magnetic intensity (B)	1 gauss	$= 10^{-4}$ Wb m^{-2}

magnetic moment	1 emu	or	$= 4\pi \times 10^{-10}$ Wb m
			$= 10^3$ A m^2

magnetisation (moment/unit volume)	1 emu	or	$= 4\pi \times 10^{-4}$ Wb m^{-2}
			$= 10^3$ A m^{-1}

It is probable that the induction interpretation, A m^2, will be adopted for palaeomagnetic and geomagnetic studies.

1

Introduction

1.1 Introduction

Rocks containing magnetic minerals become magnetised during their formation, irrespective of the age of the rock, and the study of their present
magnetic properties frequently allows this original component of magnetisation to be isolated. Measurements of this original remanent magnetisation
can be used to determine the nature of the ancient geomagnetic field and are
the only geophysical observations which allow a detailed examination of a
physical property of the Earth throughout geological time. As the geomagnetic field originates within the Earth's core, these studies are critical to
the origin and evolution of both the field itself and the Earth's interior. Such
geomagnetic studies also have a variety of geological applications, allowing
rocks to be dated and their past spatial relationship to be determined.

1.2 The early history of palaeomagnetism

The early history of palaeomagnetism is closely linked to the discovery of
the directional properties of lodestone (magnetite-rich rock). Although the
attraction and repulsion properties of pieces of lodestone were probably well
known in prehistoric times, when such attributes were considered magical, it
is generally believed that the Chinese were the first to discover the
directional properties, almost certainly several centuries B.C. The first
definite records in China date from the 1st century A.D. and magnetic
declination was certainly known there by 720 A.D., when comparisons were
made between geographic south and the direction of the south-seeking
magnetic compass needle.

In Europe the earliest known reference to the directional properties of

1

lodestone is the description by Alexander Neckham in 1190 of a fairly advanced form of compass which was apparently well known at that time. The dipole nature of a magnet, with north-seeking and south-seeking poles as specific points on its surface, was discovered in Europe by Petrus Peregrinus in 1269, and he also related the direction of a compass needle to the rotation poles of the Earth, then believed to be the axis of the universe. The independent European discovery of magnetic declination was probably made at roughly the same time, or slightly later, although such variations were largely attributed to non-uniformity of the lodestone until the end of the 15th century. The inclination of a suspended magnetic needle or lodestone from the horizontal was certainly known to the Chinese but appears to have been found independently in Europe by Peregrinus and subsequently rediscovered on two separate occasions in 1544 by Georg Hartmann and in 1576 by Robert Norman. The variation in declination from one region to another does not appear to have been observed until the mid 16th century when João de Castro made 43 observations of declination between 1538 and 1541 while voyaging around Africa to the Red Sea. In 1546 the Flemish cartographer, Gerhard Mercator, noted similar discrepancies between geographical and compass coordinates in various other parts of the world.

William Gilbert, often considered to be the first true scientist, made use of these observations, together with the work of Peregrinus and others, in his treatise, 'De Magnete' published in 1600, in which he described the Earth's magnetic field in terms of a uniformly magnetised sphere, a terella. Subsequently, in 1635, Henry Gellibrand found significant differences between the declinations measured in London in 1580, 1622 and 1634, and concluded that declination varied not only regionally but also with time. This time-change in the direction of the Earth's field is known as its secular variation.

Large scale charts of the geomagnetic field were compiled by Edmund Halley, and showed lines of equal declination in the North and South Atlantic for 1700, and the first charts of equal inclination were published in 1768 by Johann Carl Wilcke. At this time it was generally thought that the longitude of a ship could be determined from such charts, although the improvement in chronometers reduced the interest in this application. Nevertheless, there were sufficient data available for van Bemmelen, in 1899, to construct maps of declination values at 50 year intervals from 1550 to

1700. There were also sufficient data for Gauss, in 1839, to utilise his newly developed spherical harmonic analysis to demonstrate mathematically the essentially dipolar nature and internal origin of the Earth's magnetic field.

Studies of the magnetisation of rocks must, of course, go back to the original discovery of lodestone and the use of the compass as a navigational device must have been accompanied by the discovery that many rocks were themselves sufficiently magnetic to deflect a compass needle. Most of these strongly magnetic rocks have been magnetised by lightning and Alexander von Humboldt in 1797 gave this explanation for the variation in compass readings on the summit of a mountain in the Palatinate. Improvements in magnets and compass mountings meant that by the mid 19th century it became possible to detect the much weaker remanent magnetisation of igneous rocks which had not been struck by lightning, and Delesse, in 1849, showed that some lavas were uniformly magnetised parallel to the Earth's magnetic field.

In 1853 Melloni found that certain Italian lavas, on Mt. Vesuvius and in the Phlegraean Fields, possessed a remanent magnetisation and, working with Forstermann, in 1859 showed that this magnetisation was lost on heating the lavas to 100°C, but was regained on cooling. This work was extended in 1894 and 1895 when Folgerhaiter found that the directions of remanent magnetisation in both lavas and baked pottery was definitely associated with the direction of the Earth's magnetic field at the time the materials were heated and cooled, and that this original direction could be preserved for at least 2000 years. Similar magnetic stability was suggested by Brunhes and David in 1901 for igneous rocks when they found that lavas and baked soils in contact with them had identical directions. David, in 1904, found that blocks of lava used to construct a temple in the 1st century A.D. had identical inclination to lava in the quarry from which they were taken, so that the directions of remanent magnetisation of the lava had been preserved unchanged during removal, transportation and building of the temple.

Rocks of opposite polarity to the present Earth's field direction were reported from India in 1860 by Broun, but the first observation of reversed magnetisation which was definitely not associated with anomalous effects of lightning was by Brunhes in France in 1906. Subsequently reversely magnetised rocks were found in Spitsbergen, Greenland and Australia by

Mercanton between 1910 and 1926, confirming the world-wide nature of the phenomenon. Further confirmation of this and the identification of a reversed polarity period in the early Quaternary (*c.* 1 million years ago) came from studies by Matuyama in 1929 of rock samples from Japan, Korea and Manchuria.

The pioneer of modern palaeomagnetic studies was Raymond Chevallier whose work on the historical lavas of Mt. Etna was published in 1924 and 1925. This work covered the uniformity of magnetisation within individual flows, the distinction of zones affected by lightning and the determination of variations in the direction of the Earth's magnetic field since the 12th century. Chevallier's geomagnetic determinations based on the remanence of rocks were substantiated by comparison with contemporary historical records and similar agreement between observatory and palaeomagnetic determinations have now been made in many parts of the world.

Thus, by 1930 most of the fundamental discoveries about palaeo-magnetism had been made; subsequent work has entailed the development of more sophisticated techniques of measurement and analysis, making possible the precise examination of the magnetisation of rocks several million times weaker than a normal bar magnet. Such improvements could only come with the increased knowledge of the physical processes involved, largely due to the work of Louis Néel, and the development of statistical techniques by Sir Ronald Fisher. These investigations were largely stimulated by the possible applications of palaeomagnetic techniques to various geological and geo-physical problems. Many people have contributed to more recent develop-ments and it becomes invidious to specify individuals. In the main, however, the major results have been obtained by only a dozen or so research workers who, despite the previous scepticism of the majority of scientists, neverthe-less continued to develop the subject to its present status.

2

The physical basis

2.1 Introduction

In this chapter magnetic concepts are outlined from an atomic basis to the behaviour of large solids, such as rocks. The methods of magnetisation are discussed in relation to the ways in which rocks could become magnetised naturally. or in the laboratory, and the effect of time on such processes is particularly emphasised. Finally the ability of magnetic materials to acquire a magnetisation precisely aligned with the original ambient field is discussed.

2.2 Magnetism on an atomic scale

Electromagnetic theories are based on the fundamental observation that a magnetic field is produced when an electrically charged particle is in motion. An electron, a negatively charged particle, spinning about its own axis, has a dipole magnetic field similar to that of a bar magnet, i.e. with a north and south pole. An atom, therefore, possesses a magnetisation due to the spin of its electrons and in addition has a magnetisation caused by the motion of the electrons as they orbit its nucleus (Figure 2.1), but this orbital magnetisation is lost in a solid as a result of the interaction of magnetic fields of neighbouring orbits. Within the atom, the orbiting electrons form shells at different energy levels, inside which the electrons pair off their opposing spins, thereby cancelling their spin magnetisation. However, unpaired electrons can exist within different shells so that several unpaired electrons, with their uncancelled spin magnetisations, can exist in one atom. The magnetic moment caused by the spin of each unpaired electron is termed a Bohr magneton (μ_B), given by:—

$$1 \, \mu_B = M_{spin} = \frac{e \, h}{4\pi mc} = 9.274 \times 10^{-21} \, G$$

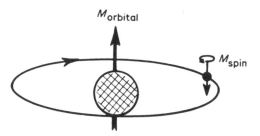

Figure 2.1   The magnetism of an atom
Magnetisation arises from the spin of an electron about its axis and from the orbital
motion of the electron about its nucleus.

where e is the charge and m is the mass of an electron, h is Planck's constant
and c is the velocity of light.

Since all substances contain electrons they respond to an applied magnetic
field. The relationship between the induced magnetisation (M) and the strength
of the applied field (H) is given by the absolute *susceptibility* ($\chi = M/H$). The
measurement of magnetic susceptibility can be used to distinguish between
substances which do or do not contain unpaired electrons. When a field is
applied to a substance *without unpaired* electrons, the electrons orbital paths
rotate, thereby creating a moment in opposition (antiparallel) to the applied
field. However, this induced moment is lost as soon as the applied field is
removed. Behaviour of this type, which is independent of temperature, is
termed *diamagnetic* and the diamagnetic susceptibility is given by:—

$$\chi_{dia} = -N \frac{e^2}{6mc^2} \Sigma \bar{r}_i^2$$

where N is the number of atoms per cubic cm and $\bar{r}_i^2$ is the mean square
distance of the ith electron from the nucleus. A magnetic field applied to a
substance *containing unpaired electrons* rotates their spin magnetisations
parallel to itself. However, this alignment is opposed by thermal fluctuations
of the individual atoms so that they become randomised when the field is
removed. This parallel behaviour is termed *paramagnetic* and the tempera-
ture-dependent paramagnetic susceptibility is given by:—

$$\chi_{para} = \frac{N\mu^2}{3kT}$$

where N is the number of atoms with unpaired electrons per cubic cm, μ is the average number of Bohr magnetons per paramagnetic atom, k is Boltzmann's constant and T the absolute temperature. The magnetic properties of both diamagnetic and paramagnetic substances are weak, with susceptibilities of about 10^{-6} G Oe^{-1}. However, a few substances (e.g. Fe, Co, Ni) contain unpaired electrons which are magnetically coupled between neighbouring atoms. This interaction results in a strong *spontaneous magnetisation*, i.e. in the absence of an external field, and in the remarkable property of being able to retain the alignment imparted by an applied field after it has been removed. These properties are several orders of magnitude greater and completely dominate those of diamagnetic or paramagnetic atoms in the same substance. This type of behaviour is termed *ferromagnetic* and is the behaviour referred to in everyday terms as 'magnetic.'

In ferromagnetic substances there are two types of quantum mechanical coupling between unpaired electrons. *Direct exchange* coupling occurs when the electron spins are linked between two adjacent atoms; when the separation of the atoms is greater than the diameter of their electron paths the exchange is negative but when they are closer together the exchange is positive. A *super-exchange* coupling occurs when the interaction between two paramagnetic atoms takes place through an intermediate atom, such as oxygen. This transmits a negative exchange between the two paramagnetic atoms by means of electrons which are shared between them. Both types of exchange interaction may be present in a crystal lattice so that its ferromagnetic characteristics are the sum of different interactions and depends on the size, distribution and relationship of the paramagnetic atoms. As the quantum-mechanical forces causing the spontaneous alignment are opposed by thermal fluctuations of the electrons both interactions are temperature dependent, so that, on heating, the ferromagnetic properties are lost at a certain temperature, the *Curie temperature* (T_c), which is specific to the geometry and composition of the particular crystal lattice. On heating above the Curie temperature, a ferromagnetic material still retains its paramagnetic atoms, but these are no longer coupled and it therefore behaves as a simple paramagnetic substance.

Ferromagnetism, in the broad sense of the term, can be divided into three types of behaviour (Figure 2.2). The simplest occurs in iron, nickel, cobalt and alloys of these metals where the exchange interactions result in the

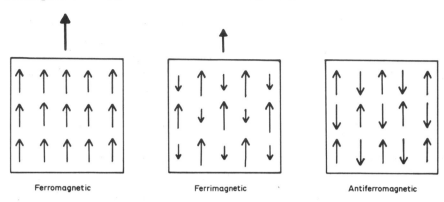

Figure 2.2   Ferromagnetism, ferrimagnetism and antiferromagnetism
In small volumes of ferromagnetic substances, all magnetic dipoles are parallel and there is a correspondingly large magnetisation. In ferrimagnetic substances, the magnetic dipoles are antiparallel, but one lattice dominates the other so that it has a low magnetisation. In antiferromagnetic materials, both antiparallel lattices are equal and the substance has no external magnetic field.

parallel coupling of all internal magnetic dipoles associated with unpaired electron spins. This results in the acquisition of a very strong magnetisation, the direction of which readily follows directional changes of the applied field. This behaviour is termed *ferromagnetic* in a strict sense. The oxides and other compounds of these metals usually have a more complex behaviour as their internal magnetic dipoles are coupled by superexchange and are therefore antiparallel to each other. The crystal lattice of these substances can be divided into two sublattices which may, or may not, be exactly the same in composition and form. Each sublattice contains parallel magnetic dipoles, but the two sublattices are antiparallel to each other. When one sublattice contains more dipoles than the other, the substance exhibits ferromagnetic behaviour due to this magnetic imbalance, but the net magnetisation is weaker than that of a (strict sense) ferromagnetic material. This type of magnetic behaviour is termed *ferrimagnetic*. When the two sublattices are crystallographically identical, the numbers of magnetic dipoles in each antiparallel sublattice are equal and so cancel each other out. The magnetic behaviour of such materials is termed *antiferromagnetic*. In practice, substances which are theoretically antiferromagnetic often exhibit a small residual, spontaneous magnetisation which arises because the lattices are rarely perfect, as they contain dislocations, vacancies, etc., even in synthetic materials, and the two sublattices are not perfectly antiparallel; such magnetisation is usually referred to as *parasitic ferromagnetism*.

Most naturally occurring magnetic minerals are either ferrimagnetic or imperfect antiferromagnetic in their behaviour, although strict ferromagnetism is common in extraterrestrial rocks, meteorites and lunar samples which contain large amounts of iron and nickel alloys. (Subsequent discussion will use the term ferromagnetism in its broad sense, unless specifically stated otherwise.) The magnetic properties of individual minerals are discussed in Chapter 3.

2.3   Domains and macroscopic magnetisation

The quantum-mechanical forces within a crystal, magnetocrystalline forces, regulate the pattern of its atoms and obviously determine the location and orientation of its paramagnetic atoms. In such an ordered structure there are certain 'easy' directions of minimum energy along which the magnetic dipoles prefer to be aligned. This alignment results in the formation of separate north and south poles ('free' poles) at the surface of the crystal although this separation is opposed by the mutual magnetostatic attraction of the poles. This conflict between the magnetocrystalline and magnetostatic forces is resolved by the development of small volume *domains* within which the individual dipoles are spontaneously aligned, which are separated from each other by narrow zones, called *Bloch walls*. Within these walls the individual dipoles gradually cant over from the direction of one domain to that of the next (Figure 2.3a). (In magnetite, the domains are of the order of 10^{-5} cm in diameter with walls some 10^{-6} cm thick; in haematite the domains are up to 10^{-3} cm in diameter, but in both minerals, the dimensions are controlled by the shape and size of the grains and the presence of imperfections in the lattice.) In the absence of any external influences, these domains arrange themselves with their north and south poles as close to each other as possible thus reducing the magnetostatic energy. This is best achieved when 'closure' patterns are formed (Figure 2.4a) so that the substance has no net external magnetic field, although this 'non-magnetic' state is radically different to that existing in a substance above its Curie temperature when all spontaneous magnetisation is

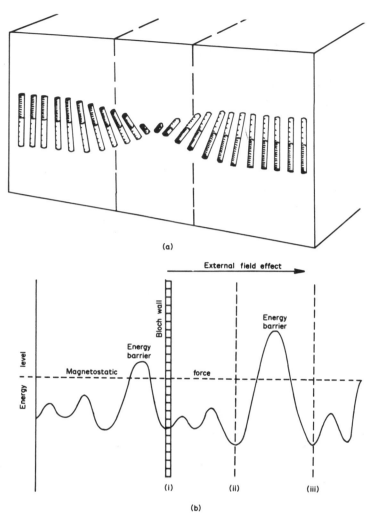

Figure 2.3   The Bloch wall

(a) The individual magnetic dipoles gradually cant over from their spontaneous direction in one domain to that of the bordering domain. (b) When an external field is applied, the Bloch wall 'unrolls' from (i) to (ii) crossing small energy barriers within the crystal framework and is able to return to position (i) when the field is removed. When a stronger field is applied, the wall is 'unrolled' as far as position (iii), past a major energy barrier, and is unable to return to its previous position. Over prolonged periods of time, however, the magnetostatic force continues to try to return the wall to its previous minimum energy position and thermal activation gradually allows individual atoms to pass through the barrier.

10

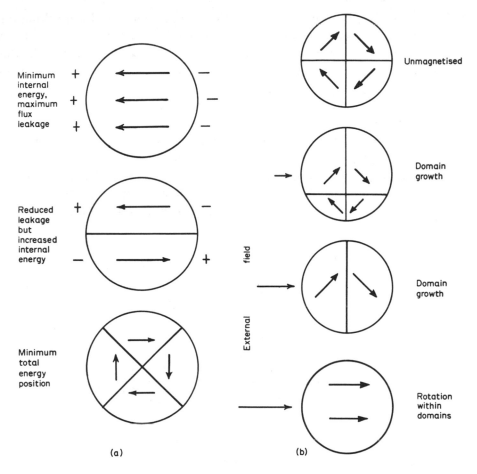

Figure 2.4   Domain patterns

(a) Alignment of magnetic dipoles within the particle creates an external field which attempts to demagnetise the particle. In order to reduce this external field and still have as many dipole aligned as possible, the particle becomes subdivided into domains so that, in its demagnetised state, the domains form closed loops with no external field. (b) When a field is applied to a 'non-magnetic' particle, domains in which the magnetisation is parallel or has a component of spontaneous magnetisation parallel to the applied field grow by unrolling their domain walls. At high fields, some directions of spontaneous magnetisation 'jump' into approximate alignment with the applied field, but it is not until very high fields are applied that the individual directions within a domain are forced into alignment.

lost. When a weak magnetic field is applied to a system of closure domains those parallel to the applied field grow by 'unrolling' their Bloch walls at the expense of non-aligned domains, but when the field is removed, the magnetostatic forces are able to return the wall to its previous position (Figure 2.3b). A stronger applied field may move a domain wall through various energy barriers within the crystal lattice, formed by lattice imperfections, impurities, etc., and after the field is removed the magneto-static forces may be incapable of moving the wall back to its original position. The material then retains an *isothermal remanent magnetisation* (I.R.M.) associated with domains which remain enlarged (Figure 2.4b). If much stronger magnetic fields are applied, then at some particular field (H_{sat}), all possible domain wall movements will have taken place across all

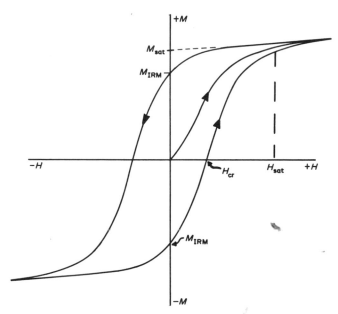

Figure 2.5   The hysteresis loop

As a magnetic field H is applied to an unmagnetised substance (i.e. its domains form closed loops), the initial increase in its magnetisation (M) gives its susceptibility ($\chi = M/H$). As higher fields are applied, its intensity of magnetisation approaches asymptotically its saturation value (M_{sat}) which is attained in a field (H_{sat}). When the field is reversed, the intensity of the magnetisation decreases, reaching zero when the applied reversed field has a strength equal to the coercivity of the material (H_{cr}).

energy barriers so that the material is magnetically saturated ($M = M_{sat}$). At even higher fields the individual electron spin dipoles within the domains are forced into alignment with the applied field. The energy required for this rotation of the dipoles is very much greater than that required to move domain walls and, on removal of the high applied field, the specimen returns to its saturated value. A graph of the induced magnetisation against the strength of a reversed applied field (Figure 2.5) shows a loop, the *hysteresis loop*, on which most of the fundamental magnetic properties can be defined.

In fine-grained material, the particles may be so small that the separation of 'free' poles may be insufficient for the magnetostatic energy to establish domain walls. In the resultant 'single-domain' grains, only high field individual dipole rotation is possible. A high field applied at an angle to the 'easy' direction of a single domain particle will rotate the dipoles into alignment but they relax back to their 'easy' direction after removal of the field. A high field applied antiparallel to dipoles in the 'easy' direction

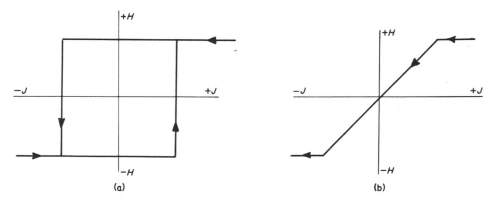

Figure 2.6   Hysteresis of a single-domain particle
When a field is applied to a single domain particle, as no domain wall movement can take place, only magnetic dipole rotation can occur. (a) A field applied along the 'easy' axis causes reversal of the direction of spontaneous magnetisation only when the saturation field is applied, so that the hysteresis loop is square. (b) When the field is applied at an angle to the 'easy' direction there is an immediate relaxation on removal of the field so that there is no hysteresis. When a field is applied to a randomly oriented group of single domain particles, the hysteretic behaviour of the material is, of course, the sum of all the individual domains and has a hysteresis loop similar to that of multidomain particles (Figure 2.5). H is the strength of the applied field and J is the intensity of magnetisation of the particle.

reverses their direction and the new reversed direction is retained when the field is removed. Such a particle has a square hysteresis loop when the field is applied along the 'easy' direction (Figure 2.6a). This single domain behaviour is particularly important in rocks which frequently contain fine-grained material upon which the weak geomagnetic field can have no effect, so that their magnetisation can remain stable for prolonged periods of time.

2.4   The effect of time, temperature and volume changes

Temperature is a measure of the average rate of vibration of atoms in a substance so that, at any one temperature, a few atoms are highly agitated. This means that, at constant temperature, the magnetisation of a few grains becomes free to rotate into different energy directions. Over prolonged periods, therefore, the magnetisation in a crystal gradually relaxes towards an equilibrium value along the direction of lowest energy, the 'easy' direction. The electrons in delicately balanced energy levels will be more readily affected by this thermal agitation, so the electron dipoles within the Bloch walls are preferentially affected, with the result that the relaxation is seen as a gradual 'unrolling' of the domain walls. In the absence of an applied field, this unrolling results in the formation of closure domains so that any original magnetisation of the material as a whole decays with the formation of these closures. In the presence of an applied field, the relaxation takes place towards the 'easy' directions which are parallel to the applied field so that a remanence is gradually built up. This time-dependent remanence in an applied field is termed *viscous remanent magnetisation* (V.R.M.), the coercivity of which must also increase with time as the domain walls gradually unroll through energy barriers within the crystal lattice.

The rate at which viscous remanence is acquired or previous magnetisation decays clearly depends on the amount of thermal agitation, i.e. the temperature, and on the strength of the magnetocrystalline and magneto-static forces of the material. For any one particle, the time taken for the magnetisation to relax to its thermal equilibrium value and direction is termed the *relaxation time* (τ). In the absence of external forces this is given by:—

$$\frac{1}{\tau} = 2f \exp\left(-VM_sH_c/2kT\right)$$

where f is a frequency factor, M_s is the spontaneous magnetisation, H_c is the field required to reverse the direction of magnetisation along the 'easy' direction at absolute zero temperature and V, k and T are the volume, Boltzmann's constant and the absolute temperature, respectively. In the presence of an applied field ($H \approx 10-20$ oersteds):—

$$\frac{1}{\tau} = f\left(1 - \frac{|H|}{H_c}\right)^{3/2} \left(1 + \frac{|H|}{H_c}\right)^{1/2} \exp\left[-\frac{V M_s (H_c - |H|)^2}{H_c \; 2kT}\right]$$

In these expressions, the frequency factor is approximately constant ($f \approx 10^{10}$ sec) and for a given material the crystalline properties, M_S and H_C, are also approximately constant except near its Curie temperature. This means that variations in the relaxation time of any one particle are almost entirely dependent on changes in its volume and temperature ($\tau \propto \log V/T$). Very fine ferromagnetic particles, less than domain size, therefore have very short relaxation times even at normal temperatures and behave *superparamagnetically*, that is, their behaviour is paramagnetic but their intensity is typical for ferromagnetic substances. The individual particles have normal ferromagnetic intensity but very short relaxation time so that they can rapidly follow directional changes of an applied field and, on removal of the field, do not hold any remanence. Slightly larger particles have relaxation times of the order of a few minutes, so that they acquire a remanence which can be measured just after the field is removed. The volume of the particle at which this occurs is termed the *blocking volume* (V_B) as it is the volume at which the direction becomes blocked within that individual particle long enough for its measurement by conventional methods. As the relaxation time of the particle is also dependent on temperature, there is also a *blocking temperature* (T_B) for any individual particle when the relaxation time is a few minutes. This concept of a blocking temperature and blocking volume is critical to any understanding of the processes of magnetisation and demagnetisation. At a high temperature ($T > T_C$), a particle behaves paramagnetically until the temperature falls to the Curie temperature ($T = T_C$) when spontaneous magnetisation develops (Figure 2.7). At this point, the particle behaves superparamagnetically until the temperature falls to its characteristic blocking temperature ($T = T_B$) when the direction of magnetisation acquired by the particle becomes frozen or blocked in for the duration of the experiment

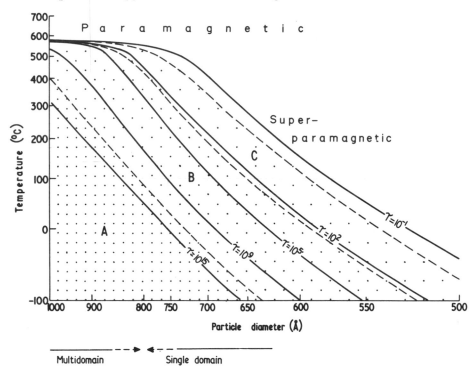

Figure 2.7   Temperature, volume and relaxation time
When either the temperature decreases or the particle volume increases, the relaxation time (τ secs.) increases. The example used in this illustration is magnetite. Region A corresponds to relaxation times comparable to geological time and region C to laboratory time.

($\tau \cong 5$ min). Further cooling reduces thermal agitation so that the corresponding exponential increase in the particle's relaxation time eventually becomes comparable to or greater than geological time scales ($\tau > 10^9$ years) when the magnetisation acquired near the blocking temperature becomes blocked indefinitely. The magnetisation acquired in this way is termed *thermo-remanent magnetisation* (TRM).

Other forms of magnetisation can be acquired in comparable ways. When the agitation, which releases the electron dipoles allowing them to relax, is not thermal, but produced by alternating magnetic fields, then a constant magnetic field over the specimen produces a magnetisation which becomes blocked in as the alternating field is reduced, producing an *anhysteritic remanent magnetisation* (ARM). Similarly, agitation can be produced by mechanical vibration and a *mechanical remanent magnetisation* (MRM) results. Alternatively, the source of agitation can be kept constant e.g. by maintaining a constant temperature, and a magnetisation produced by increasing the volume of the particle by chemical growth (Figure 2.7), so that a small particle passes from a very small, superparamagnetic state through a narrow volume range during which its relaxation time increases to millions of years, blocking in any magnetisation it carried at its critical blocking volume. Such chemical growth gives rise to a *chemical remanent magnetisation* (CRM).

The reversal of these magnetisation processes, i.e. heating the material, decreasing its volume, etc., in zero applied field reduces the relaxation time of all particles within it so that particles which already had low relaxation times at room temperature (or zero alternating field, etc.) attain their thermal equilibrium magnetisation values and orientations at moderate temperatures (alternating fields, etc.). If the temperature is then reduced, the magnetisation of particles which had attained equilibrium become blocked in their new orientations, while all others remain blocked in their previous orientations.

In natural substances, such as rocks, which contain particles of varying size, shape and composition, there is a wide spectrum of relaxation times. The remanent magnetisation observed under normal conditions is therefore the resultant of all the blocked magnetisations. Partial demagnetisation allows the particles with shortest relaxation times to reach their equilibrium values first and, in the absence of an aligning field, these are along the direction of their individual 'easy' axes. Subsequent reduction of the agitation blocks these particles in their relaxed positions so that they no longer contribute to the observed remanence. Incremental demagnetisation therefore allows the relaxation time spectrum to be investigated by the preferential removal of the remanence associated with shorter relaxation time particles, leaving remanence associated only with particles of long relaxation times which, in the case of rocks, can be thousands of millions of years.

2.5   Pressure effects and magnetic anisotropy

The application of uniaxial pressure to a ferromagnetic substance can result in an unrolling of domain walls which is reversible after moderate pressure, but may become locked by energy barriers after high pressure, resulting in a *pressure remanent magnetisation* (PRM). Hydrostatic pressure, such as that experienced by rocks during burial, appears to be capable of unrolling domain walls, but the effects are negligible and reversible in most cases, although this process has not been adequately studied for firm conclusions to be drawn. In most rocks, however, the direct effect of pressure is subordinate in its palaeomagnetic importance to that of anisotropy (see below) or thermal and chemical changes which usually accompany any significant uniaxial or hydrostatic pressure.

As it is possible for the 'easy' directions in different crystals to be aligned within a substance, it will require less energy to magnetise it parallel to the 'easy' direction than in other directions. Such *anisotropic* materials do not, therefore acquire a direction of magnetisation that is parallel to an applied field. It is critical, therefore, that the degree of magnetic anisotropy is established when determining the direction of previous fields from analyses of remanent directions. *Crystalline anisotropy* occurs when the individual crystals are not randomly oriented so that their aligned 'easy' directions give a preferred direction to the material. Magnetisation therefore results in a remanence along this direction and the remanence acquired by the material in an applied field always has a component along this preferred direction. Crystalline anisotropy is particularly temperature dependent and, in most rocks, is not significant at temperatures in excess of 300°C. In some minerals the crystalline behaviour at low temperatures is complicated either by their transition from one crystal form to another or by changes of their 'easy' directions; magnetite changes from its normal cubic form to an ortho-rhombic structure at about − 160°C and in haematite the 'easy' axes change from axial to basal plane orientations below −15°C. These *transition temperatures* mark major changes in both the magnitude and direction of crystalline anisotropy. Intrinsic *magnetostrictive anisotropy* occurs when the energy barriers within the individual crystals are not randomly oriented so that domain wall movement is restricted in certain directions. As the location of energy barriers is mainly dependent on the crystal form, this anisotropy is usually only significant when there is a crystalline anisotropy present. In most substances with aligned elongated particles, the crystalline anisotropy is subsidiary to *shape anisotropy*. Each particle develops separate poles as it is magnetised and, as in the case of domain formation, this process is opposed by the attraction of the two poles which therefore create a demagnetising field within the particle itself, the strength of which varies with the separation of the poles (Figure 2.8). These materials have a shape anisotropy because their elongated particles are more easily magnetised along their long axes and therefore a material containing aligned, elongated particles has a preferred direction of magnetisation and acquires a com-

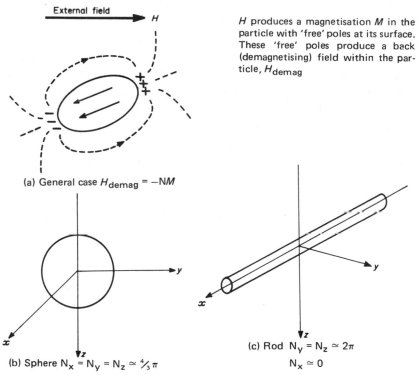

External field

H produces a magnetisation M in the particle with 'free' poles at its surface. These 'free' poles produce a back (demagnetising) field within the particle, H_{demag}

(a) General case $H_{demag} = -NM$

(b) Sphere $N_x = N_y = N_z \simeq \frac{4}{3}\pi$

(c) Rod $N_y = N_z \simeq 2\pi$
$N_x \simeq 0$

Figure 2.8   Shape anisotropy
An applied field results in a back (demagnetising) field being set up within the material which is a function of its composition and shape. The demagnetising factor (N) is shown for (a) the general case, (b) a sphere and (c) a rod.

ponent of remanence along this direction. As the shape of the particles is unaffected by temperature changes, this form of anisotropy is particularly important in substances which acquire their remanence thermally.

Induced anisotropy arises from the influence of external as opposed to intrinsic effects. Hydrostatic pressure reduces the distance between different atoms, thereby affecting the magnitude, but not the direction of anisotropy, and removal of the pressure allows the atoms to recover, so that the previous anisotropy returns. Uniaxial pressure has more complex effects. Weak pressures may distort the crystal lattice, but the distortion is reversible so that the effect on the crystalline anisotropy is also reversible. Stronger or more prolonged pressure results in permanent deformation and therefore a lattice deformation in a specific direction which has an associated crystalline and magnetostrictive anisotropy, usually perpendicular to the direction of uniaxial pressure. This deformation may also result in changes in the shapes of the particles and if, as in some rocks, the pressure is accompanied by thermal and chemical changes, there may also be some preferential growth of particles in directions perpendicular to the stress and the material therefore develops a marked shape anisotropy. A further anisotropy may occur when two or more types of ferromagnetic material are present with different Curie temperatures. On cooling the lower Curie temperature material acquires a magnetisation related to both the applied field and the field of the already magnetised higher Curie temperature material. This can eventually lead to an actual reversal of the remanent direction (Section 8.2), but more commonly results in a distortion of the remanence directions carried by the lower Curie temperature material and therefore of the substance as a whole.

3

The commoner magnetic minerals and their identification

3.1 Introduction

Many minerals contain magnetic elements, particularly iron, but the majority of these minerals have Curie temperatures which are below $0°C$ and are of no direct interest in palaeomagnetic studies. Some substances with higher Curie temperatures, such as free iron and nickel-iron alloys, are extremely rare in terrestrial rocks and are only important in studies of meteorites or other extraterrestrial rocks. This chapter is therefore mainly concerned with the iron-titanium oxides which are found in most igneous rocks and also as detrital particles in sediments. The weathered products from pre-existing rocks include the hydroxides of these minerals and silicates which on, or subsequent to, deposition give rise to the magnetic properties of sedimentary rocks. Various methods of determining the composition and physical relationship of these minerals are briefly assessed.

3.2 The iron-titanium oxides in igneous and metamorphic rocks

The most common accessory mineral in igneous rocks is 'magnetite' which often forms up to 5% of the total rock. 'Magnetite' is a loose petrological term for an opaque particle with a steely reflection and usually, but not always, seems to have crystallised before the major constituents of the rock. Under high magnification, these 'magnetite' particles can be seen to be complex, often formed of two or more intergrown minerals, although their composition is chemically simple being mostly iron, titanium and oxygen in various ratios (Figure 3.1).

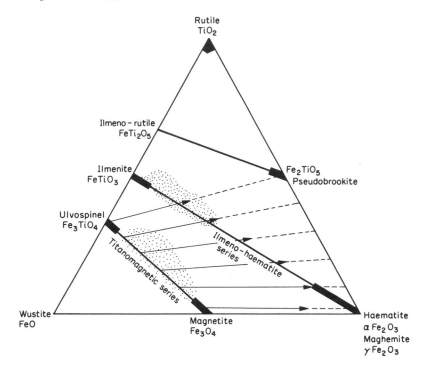

Figure 3.1   Ternary diagram of the composition of the common iron-titanium oxides
The average chemical composition of the iron-titanium oxides found in most igneous rocks is shown dotted. This actual composition can only occur in the solid solutions magnetite-ulvospinel (cubic structure) and ilmenite-haematite (rhombohedral), or as partially oxidised solid-solutions. (The direction of increasing oxidation with constant Fe:Ti ratio is shown arrowed.) In many igneous rocks, the average composition is that of two or more different minerals, the commoner compositions being shown as solid lines on the diagram. It is essential, in palaeomagnetic studies, to determine both the chemical and physical relationship of these different minerals.

In basic igneous rocks (iron-magnesium rich), the particles usually have an average composition between ulvospinel (also known as ulvite, $Fe_2 TiO_4$) and true magnetite ($Fe_3 O_4$). (The term 'magnetite' is subsequently used to refer only to true magnetite.) Both of these minerals have a cubic (spinel) crystal structure which allows a complete solid solution, the titanomagnetite series, to exist between them (Figure 3.2). Their crystal structure is simple, consisting of alternating layers of anions and cations; the unit cell

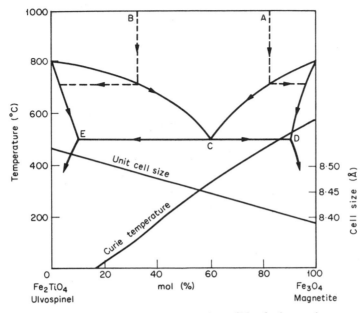

Figure 3.2   The titanomagnetite solid-solution series
The variation in Curie temperature and cell size is shown for solid-solutions of this series. A slow cooling solid-solution usually exsolves into magnetite and ulvospinel. Composition A, for example, cools to about 700°C when almost pure magnetite exsolves and the composition of the remaining solid solution changes towards composition C. At approximately 500°C, crystals of composition D and E come out of solution. Similarly, an initial composition B, exsolves ulvospinel first, and then ulvospinel and magnetite.

comprising 32 oxygen anions and 24 cations, of which 8 are in four-fold coordination (forming a tetrahedral lattice – the A lattice) and 16 are in six-fold coordination (forming an octahedral B lattice). In magnetite, Fe^{3+} cations fill the A lattice and are balanced by a similar number in the antiparallel B lattice so that magnetite is ferrimagnetic due to the unbalanced Fe^{2+} cations in its B lattice. The change in composition from magnetite towards ulvospinel is marked by an increase in size and distortion of the cell as Ti^{4+} enters the B lattice and Fe^{2+} gradually replaces all Fe^{3+}, although preferentially entering the B lattice. This means that ulvospinel itself is antiferromagnetic as it has equal amounts of Fe^{2+} in both lattices. However, the Curie temperature of the ferrite solid solution falls to below room temperature before the ulvospinel composition is approached.

The magnetic properties of the titanomagnetite series have been established using mainly synthetic minerals. The presence of greater strains, imperfections and impurities in natural minerals (Al, V and Cr often replace Fe^{3+} and Fe^{2+} is often replaced by Mg, Mn and Ca) generally slightly lowers their Curie temperatures and also somewhat reduces their saturation moments; the value for pure magnetite being 92 G/gm.

Titanomagnetite particles are often preserved in basic rocks when they are rapidly cooled if their composition is close to either magnetite or ulvospinel. Ulvospinel, although more common than previously thought, is still relatively rare, suggesting that most basic igneous rocks have an original titanomagnetite composition close to that of magnetite. In more slowly cooled rocks, however, the intermediate compositions usually exsolve from the solid solution forming intergrown crystals of magnetite and ulvospinel along the [111] crystallographic planes of the previous titanomagnetite particle. Petrological studies of these intergrowths suggest that the majority of this exsolution takes place while the rock is still cooling, but exsolution can also take place very gradually at normal temperatures. The timing of this exsolution is of fundamental importance to palaeomagnetic studies (Section 4.2) since it results in the acquisition of a chemical remanence (Section 2.4).

More acidic rocks (richer in silica) contain more oxygen in their original molten state, and therefore the mineral particles which crystallise in them have a higher oxidation state, and a composition (Figure 3.1) lying between that of the titanomagnetite series and the ilmeno-haematite series. Again, in basic rocks, the titanomagnetites may react with residual, oxygen rich, liquids so that their composition changes towards the ilmeno-haematite series. This change is not fully understood but appears to take place initially by the titanomagnetite taking up more oxygen, but retaining its cubic form by developing vacancies among its cations so that, for example, magnetite (Fe_3O_4) can oxidise at low temperatures to maghemite (γFe_2O_3), particularly when the original particle is small or its lattice structure imperfect. At some stage, however, exsolution generally takes place from these intermediate compositions, usually into magnetite and ilmenite ($FeTiO_3$), as the two end members of this series have different crystal structures and there cannot, therefore, be a continuous solid-solution between them.

In more oxygen-rich rocks, the iron-titanium oxides belong to the

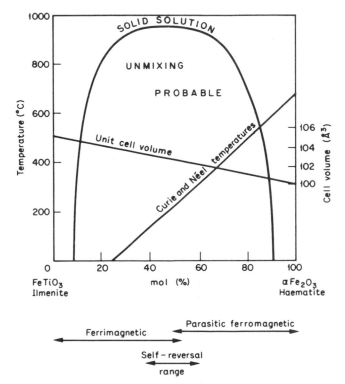

Figure 3.3   The ilmeno-haematite series

The cell size and Curie temperature are shown for solid-solutions of this series. Intermediate solid-solution compositions are rare in rocks, as exsolution usually takes place for these compositions below $900°C$, i.e. along the curve (solidus) shown, unless the original solid-solution is very rapidly cooled (see exsolution and compositional behaviour of other solid solutions outlined in Figure 3.2).

ilmenohaematite series (Figure 3.3). This series also form solid-solutions at high temperatures ($>1000°C$) since both ilmenite ($FeTiO_3$) and haematite (αFe_2O_3) have rhomboheral (corundum) crystal structures. On cooling, however, intermediate compositions almost always exsolve into intergrowths of ilmenite and haematite. (The properties of solid-solutions of intermediate compositions are uncertain, but this is not palaeomagnetically important as such natural particles are rare. The properties are discussed further in relationship to the self reversal of magnetic directions in Section 8.2.) Both ilmenite and haematite have the same amount of iron in each antiparallel crystal lattice and are therefore basically antiferromagnetic. Haematite, however, is extremely important in palaeomagnetic studies as it carries a weak, but very hard, parasitic ferromagnetism (Section 2.2). The precise origin of this parasitic ferromagnetism is uncertain, although it seems probable that it is due to either the haematite sublattices being not exactly antiparallel or the magnetisation of one sublattice being preferentially reduced by impurities, defects, etc. The sources of parasitic ferromagnetism probably vary from one crystal to the next so that only generalisations about the magnetic properties of haematite are possible. Its saturation magnetisation is approximately 0.5 G/gm and on cooling below $-15°C$, the crystalline anisotropy of pure crystals changes orientation from the basal plane of the crystal to its c axis. However, in naturally occurring haematites this 'Morin transition' may be completely suppressed.

Haematite, which can originate from the oxidation of titanomagnetites or the exsolution of ilmeno-haematites, may also appear in igneous rocks from the decomposition of other normally paramagnetic, iron-bearing minerals such as olivines, pyroxenes and amphiboles. This decomposition is mostly associated with the weathering of the rock, but can also occur spontaneously over very prolonged periods of time and is not uncommon in unweathered igneous rocks older than some 1,000 million years.

All of the above mentioned iron-titanium oxides may occur in metamorphic rocks and it is impossible to make generalisations about their occurrence. Slightly increased temperatures generally increase exsolution from any pre-existing solid-solutions, but this may also be accompanied by either oxidation or reduction. At higher temperatures, unmixed phases may remix into solid solutions which will eventually have physical and chemical properties dependent on their reactions with circulating fluids and gases, and their rate of cooling.

In both igneous and metamorphic rocks, very high oxidation from either reactions with residual fluids or, more commonly, weathering, may cause ilmenite to disintegrate into pseudobrookite ($Fe_2 TiO_5$), rutile (TiO_2) and other complex, usually amorphous, iron-titanium oxides. (Pseudobrookite and maghemite are particularly useful as they are only stable at low temperatures and are therefore diagnostic of low temperature exsolution and low temperature chemical remanence.) Most other iron-bearing minerals become oxidised to haematite, but in the presence of water, even haematite may break down, forming iron hydroxides, such as goethite (γFeO.OH) which is antiferromagnetic but carries a parasitic ferromagnetism. Yet further hydration results in the formation of various amorphous iron hydroxides, collectively termed limonite (FeO.OHnH$_2$O).

3.3 The magnetic minerals in sedimentary rocks

Sedimentary rocks are formed of varying amounts of detrital particles and biochemical deposits, usually in a calcareous or siliceous cement. On the whole, the biochemical deposits contain little magnetic material, although some carbonaceous deposits contain iron sulphides (Section 3.4). The detrital particles, may include initially, any iron-bearing minerals derived from pre-existing igneous, metamorphic or sedimentary rocks. Most of these are chemically unstable in air and oxidise, many of them forming haematite. The reaction of these oxides, particularly haematite, with water results in the formation of iron hydroxides and clay particles. One such iron hydroxide, goethite, has magnetic properties at room temperature, but most are non-magnetic.

The mineral content of the detrital particles forming a sediment clearly depends on the nature of the rocks eroded, and the types and rates of erosion, transport and deposition. During all of these processes, there is the general formation of haematite by oxidation and its hydration to form iron-hydroxides. These continue to form after deposition until the weight of overlying sediments compacts the lower sediment, and dries it out. This drying out dehydrates the clays and iron-hydroxides, usually resulting in the formation of haematite. All of these diagenetic processes, therefore, tend towards the formation of haematite within consolidated sedimentary rocks, although a cessation of chemical activity at any stage may leave a complex mineral assemblage. The compaction and drying out is usually accompanied by a cementation of the detrital particles, usually by carbonates and sometimes by silica which crystallise out of circulating fluids. These cements often contain haematite in a very finely dispersed form, commonly derived from within the sediment itself but it may also be brought in from outside, for example, by streams eroding highly oxidised rocks.

The formation of haematite in sediments is of fundamental importance to palaeomagnetic studies as it is accompanied by the acquisition of a chemical remanence (Section 2.4). Studies of recent sediments suggests that this process may take many thousands of years. On the other hand, red siltstones, which have been found to be most useful in palaeomagnetic studies, appear to preserve evidence of past geomagnetic field changes having periodicities of 1,000 years or so (Section 8.6). This suggests that, in certain sediments at least, the process of oxidation must have been completed at, or very shortly after, the time of deposition of the particles, or alternatively any chemical remanence has adopted the direction of magnetisation of the original detrital magnetic particles.

3.4   Less common magnetic minerals

Pyrrhotite is the most important magnetic mineral which is not an iron oxide and occurs occasionally in massive basic igneous rocks. It is an iron sulphide (FeS_{1+x} with x varying from 0 to 1) intermediate in composition between troilite (FeS) and pyrite (FeS_2). Troilite is very rare in terrestrial rocks but quite common in meteoritic material. It has a monoclinic (pseudohexagonal) crystal structure and is antiferromagnetic. An increase in its sulphur content causes vacancies to appear, replacing the iron atoms, which have ordered locations so that, for certain compositions, particularly in the region of $x = 0.14$, pyrrhotite has strong magnetic properties. Its magnetic behaviour is generally considered to be that of a ferrite and this is supported by evidence of self-reversal in some natural pyrrhotites (Section 8.2). However, it has also been reported as ferrimagnetic for $x > 0.08$ and antiferromagnetic for composition $x < 0.08$. Similar uncertainties exist about the Curie temperature which is generally reported between 300 and 350°C, although 500°C has been observed. Natural pyrrhotites show considerable variation in magnetic behaviour, although they are generally considered to have a low coercivity. In most terrestrial rocks, the iron sulphides appear as pyrite (FeS_2) which has a cubic structure and is nonmagnetic at normal temperatures.

Almost all other iron-bearing minerals are paramagnetic at normal temperatures and the few minerals which are magnetic at room temperature, jacobsite ($MnFe_2O_4$), cubanite ($CuFe_2O_3$), magnesioferrite ($MgFe_2O_4$), trevorite ($NiFe_2O_4$), chromite ($FeCr_2O_4$), franklinite ($(Fe,Zn,Mn)(Fe,Mn)_2O_4$) and iron oxides incorporating rare earths (Fe_5(Rare

Earth)$_3$O$_{12}$) are sufficiently rare to be irrelevant to normal palaeomagnetic studies. Occasionally other minerals have been reported to have magnetic properties at normal temperatures, e.g. cassiterite (SnO$_2$), but their magnetisation, like that of some clays, micas, etc., is related to haematite or magnetite inclusions within their structure and not to the actual mineral itself.

Extraterrestrial rocks contain large quantities of free iron and iron-nickel alloys. These are ferromagnetic (s.s.) and therefore have a very high saturation magnetisation and low coercivity, so that their viscous magnetisation may swamp any stable remanence carried by iron and nickel oxides.

3.5   The identification of magnetic minerals

In most terrestrial rocks, the remanence of importance to palaeomagnetic studies is carried by magnetite in igneous rocks and haematite in sedimentary rocks. However, any interpretation of the remanence depends on a knowledge of the origin of the minerals carrying the remanence, and this can only come from a study of the relationship between the magnetic minerals themselves and related minerals. Various techniques are available for these investigations, using either separated particles or the total rock.

As magnetic minerals have a high density (about 5) they can be separated with a magnet from a high density fraction of crushed rock obtained by flotation in heavy liquids or by centrifuging. The magnetic extract can then be analysed although it normally requires the removal, by hand, of silicates which still adhere to the particles. Standard chemical analyses can determine the ratios of ferric and ferrous iron, titanium, etc. and information on the cell size, structure and distribution of atoms within the powder or individual crystals can be obtained by X-ray diffraction or fluorescence and Mössbauer techniques. Study of the gross magnetic properties of the rock as a whole allows some mineral identification to be made by the examination of, for example, the coercivity and saturation magnetisation (Section 2.3). More precise identification can be obtained by determining the Curie temperatures of the magnetic constituents (Figure 3.4). This can be done quickly and precisely by means of a thermomagnetic analysis, but the heating process may result in oxidation (or reduction) of the magnetic minerals themselves or of other minerals. Such

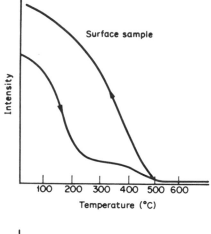

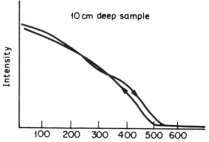

Figure 3.4   Curie temperature analysis
The lower graph shows a repeatable heating and cooling curve for a fresh sample of rock containing magnetic material with a Curie temperature of 520°C., corresponding to a magnetite composition. The upper graph shows the behaviour of a weathered, surface sample of rock, with Curie temperatures of 220°C and 500°C., probably goethite and magnetite. However, the goethite is unstable on heating and changes to magnetite so that the curve is not reproducible.

chemical changes can be detected by comparison of the susceptibility before and after heating and cooling.

The study of polished sections of a rock offers the most comprehensive information on both the composition and physical relationship of the different minerals. As the iron-titanium minerals are opaque, standard microscopic methods can be used for their identification in reflected light. The use of electron beams in the microprobe allows not only similar methods of identification but also quantitative measurement of the content of iron, titanium and most impurities. Both of these techniques allow detection but not analysis of particles as small as 0.5 microns in diameter, only just within the size range of single domain particles, but only multidomain particles, diameters of $> 5-10$ microns, can be properly studied by these methods.

New techniques are currently being developed to differentiate the magnetisation of large magnetic grains from that of finely dispersed material and to measure the remanent magnetisation of individual grains.

Present techniques are not capable of examining the properties of particles over their full size range. Single-domain particles which, normally forming less than $1-5\%$ of the total magnetic material present in a rock, have exceedingly high coercivities (several 100 Oe for titanomagnetites and in excess of 15000 Oe for haematite) and are the magnetic components which retain the remanence acquired by very ancient rocks. In general it is assumed that the compositional and physical relationships determined for multi-domain particles are the same for smaller particles.

4

The magnetisation of rocks and its physical analysis

4.1   Introduction

The application of physical theories to rocks is complicated by several factors. The magnetic components at the time of their formation are rarely chemically or physically homogeneous and their properties change with time even in the absence of later geological events. Furthermore their remanence is generally carried by only 5% or less of the total magnetic material present, and this is dispersed throughout the rock so that a rock physically corresponds to a weak 'solution' of fine-grained ferromagnetic particles within a predominantly paramagnetic matrix. In this chapter, the processes of magnetisation which occur during the formation of the rock are considered first. Most of this magnetisation takes place over a few years, in the case of igneous rocks, and a few hundred years in the case of sediments; the age of this magnetisation is therefore the same as the rocks themselves and is termed *primary*. The magnetisation now observed in the rock, the *natural remanent magnetisation* (NRM), consists of any remaining primary magnetisation together with all subsequent, *secondary*, magnetisations; the sources of these secondary magnetisation are discussed next. There are various methods of isolating any magnetisation which is sufficiently stable to have persisted since the formation of the rock. However, such stable remanence is not necessarily primary in origin and methods of determining its age are discussed. Before this primary remanence, having been isolated, can be interpreted in terms of the strength and direction of the ambient field at the time of the rock's formation, it is necessary to determine if the remanence is isotropic, i.e. if it was acquired exactly parallel to the ancient field, and whether inhomogeneity effects prevent true measurements of the remanent directions. Only a general outline of the analytical methods is given and instrumental considerations are discussed in the following chapter.

4.2   Primary magnetisation

In igneous rocks the primary magnetisation is acquired as the hot rock cools to normal surface temperatures and in sedimentary rock it is the magnetisation acquired during deposition and diagenesis (those processes which convert a wet slurry of sediment into a compact rock).

(*i*)   *Igneous rocks* have a primary magnetisation which, for most practical purposes, may be considered purely thermoremanent in origin (Section 2.4). As the hot, liquid magma cools from temperatures in excess of 1,000°C, it solidifies, usually over 800°C, and eventually reaches the Curie temperature of its ferromagnetic constituents, generally between 400 and 600°C. Most domains in basic igneous rocks have high blocking temperatures so that the prevailing field direction is blocked within the rock at about 50°C below the Curie temperature (Figure 4.1). Further cooling increases the relaxation time of the domains and the ambient field direction then becomes locked within the high blocking temperature domains for geological periods of time.

In detailed studies, this concept of thermoremanence is not precisely valid as a cooling igneous rock is simultaneously undergoing complex physico-chemical interactions. Minerals which are stable at high temperature may be unstable, physically or chemically, at lower temperatures and break down into their constituent parts (Sections 3.2 and 3.3).

Exsolution takes place at different rates, mostly while the rock is cooling, and depends particularly on the speed of cooling and on the chemical environment, especially the concentration of oxygen. In some cases it may be temporarily or even entirely inhibited. As exsolution procedes the diameter of the growing crystallites increases to beyond their blocking diameter, and, if this occurs below their Curie temperature, they acquire a stable chemical remanent magnetisation (Section 2.4). The natural primary remanence of igneous rocks is to some extent simplified by their slow cooling rates compared with laboratory experiments; submarine lavas are chilled rapidly, but subaerial lavas remain hot for several years and deep

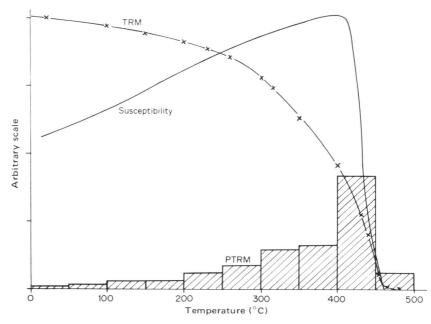

Figure 4.1   The acquisition of thermal remanence
As a rock cools from above its Curie temperature, most of its remanence is acquired within about 50°C of its Curie temperature and the total remanence at room temperature is the sum of all the remanence acquired at higher temperatures (TRM = ΣPTRM). The susceptibility (χ) also increases markedly at the Curie temperature, but then decreases with decreasing temperature.

intrusive rocks may retain their heat for hundreds and possibly thousands of years. These rocks will acquire a high temperature viscous remanence (Section 2.4) since, at temperatures 100–200°C below the Curie point, most of their domains will still be free to relax towards the ambient field direction. This slow acquisition of remanence tends to even out short term fluctuations in the local field, such as magnetic storms, lightning etc.

As the rocks cool they contract, resulting, for example, in such features as columnar jointing. Such pressures appear to have no direct effect on either the intensity or direction or remanence, as the magnetic properties are essentially functions of composition and grain dimensions determined prior to the tensional fracturing.

The complexities of behaviour in cooling rock make it difficult to predict the properties of an individual sample. In a dyke, for example, the edges tend to cool rapidly and are therefore fine grained, and if these grains are of about single-domain size, they carry a stable, mainly thermoremanent magnetisation. In the interior, which has cooled more slowly, the grains of the opaque minerals tend to be larger and multi-domained, and therefore less stable. However, exsolution is more likely to have taken place and these larger grains are often broken down into crystallographically and magnetically distinct units which can carry a stable, dominantly chemical remanent magnetisation. It is usually unimportant that the natural remanence is formed by varying amounts of thermal, chemical and high temperature viscous magnetisation, as most of the magnetisation which is stable over geological time is acquired above about 300°C and can still be related directly to the ambient field at the time of formation of the rock. However, in particular instances, a stable secondary low temperature chemical remanence may have been acquired over prolonged periods from both continuing exsolution and the breakdown of iron-bearing minerals such as olivines and pyroxenes. This stable secondary remanence generally increases in importance with the age of the rock, but may have been acquired parallel (or antiparallel) to the magnetic field of the already magnetised minerals. In most rocks, however, this secondary magnetisation is related to the local direction of the Earth's magnetic field prevailing at the times when the individual crystallites grew through their critical diameter and must be removed when isolating the primary remanence.

(ii) *Sedimentary rocks* contain magnetised particles eroded from pre-existing rocks. As these particles fall through the water, they become aligned with the ambient field and this alignment may be preserved on reaching the sediment. Deposition experiments using crushed sediments or synthetic particles show that the azimuth is preserved during deposition, even from fast flowing streams (> 5 cm/sec), but the inclination is systematically reduced by the rolling of the individual particles on hitting the sediment surface and this 'inclination error' is increased if the deposit is laid down on a sloping surface. However, if the wet sediment is left for a few hours, or very minor vibrations are introduced, the particles become aligned parallel to ambient fields as low as 0.004 Oe (the present Earth's field is about

0.5 Oe). As most recent sediments do not appear to reflect such inclination errors, it seems that any deviation introduced during deposition is lost within a few hours if the water content is greater than 50%; and a few years for lower water concentrations, by the continual action of microseisms and other seismic events. However, the situation still requires much further investigation.

The composition of the deposited particles varies according to the nature of the source rocks, the climatic conditions, and the distance from the source, and many are chemically unstable in their new environment. As the sediment dries out, the titanomagnetites oxidise and the iron-hydroxides are reduced, both tending to form haematite, which is also formed as other iron-bearing silicates, such as clays, olivines and pyroxenes, disintegrate. Many, possibly most, sediments will therefore acquire a chemical remanence as these new minerals grow. This process is normally accompanied by the circulation of various fluids which usually leave behind a cement, commonly a carbonate or silicate, which partially or completely fills the interstices between the deposited particles. As this cement frequently contains very finely divided haematite, the cement itself may acquire a chemical remanence as crystallisation proceeds. During compaction, which usually occurs at the same time as dehydration and cementation, the beds are compressed by the weight of overlying sediments. This may cause the physical rotation of some magnetic particles into the plane of the bedding, but most of the weight is taken by pressure points on the larger particles and it is unlikely that the smaller particles, or cement, which carry most of the stable remanence, are directly affected by such compression.

The diagenetic processes of converting a wet slurry into a compact rock may be prolonged so it is not surprising that few types of sediment carry a stable remanence related closely to the time of deposition. The most suitable sediments for palaeomagnetic studies of rocks older than 100 million years are red siltstones; these have mostly formed by stream deposition in desert environments and probably underwent most of their oxidation and dehydration during the first thousand years or so after deposition. Other types of sediment, particularly varves and deep sea sediments, are known to be capable of retaining a primary magnetisation and any sediments which have remained chemically inert since shortly after their formation are potential carriers of primary remanence. Most other sedimentary rock types, however, either acquire a chemical remanence over prolonged periods of time or are unable to retain a primary remanence.

4.3   Metamorphic rocks

Rocks which have been subjected to pressures or temperatures sufficiently great to cause either physical or mineralogical changes will only rarely preserve any of their primary remanence. The magnetisation of metamorphic rocks is therefore usually entirely secondary but if it takes place at a distinct, identifiable time it is useful to consider such magnetisation as primary. For example, the baking of country rock by a dyke can be related to a specific time and the thermoremanence associated with it can be considered primary. Most metamorphic activity, however, involves varying degrees of heat and pressure over prolonged periods and is accompanied by the circulation of chemically active fluids and gases.

Deep burial is usually accompanied by regional heating to a few hundred degrees and these temperatures may be maintained for millions of years, so that a high temperature viscous remanence tends to dominate in such rocks. (A temperature of some $300°C$ maintained for a few million years will allow all domains within the rock to relax and become blocked in the prevailing magnetic field as the regional temperature decreases.) The direct magnetic effects of hydrostatic pressure during burial appear to be reversible and are lost as the rock is re-exposed, but the effects of uniaxial pressure are less well established. In general they appear to be reversible when there is no plastic deformation of the magnetic minerals. However such deformation is common during the application of pressures for prolonged periods and at moderate to high temperatures. This action is frequently accompanied by chemical activity, resulting in mineral growth in the plane perpendicular to the direction of pressure and the acquisition of both a chemical remanence and anisotropic properties. In short the main effect of metamorphism is to impart a chemical and high temperature viscous remanence which is accompanied by anisotropy introduced as minerals grow under the influence of pressure. In view of these complexities few palaeomagnetic studies have been made of metamorphic rocks, but there appears to be a potential use of these anisotropic properties as a measure of the metamorphic conditions to which the rocks have been subjected (Section 9.5).

4.4   Secondary magnetisation

The effects of metamorphic activity, prolonged exsolution and diagenesis are strictly secondary, but these are not considered in this section, which concerns the effects of low temperature viscous remanence, lightning and weathering.

A viscous remanence grows as the age of a rock increases and more domains become free to relax into the later geomagnetic field (Section 2.4). This means that in old rocks, much of the primary magnetisation will have decayed and secondary viscous magnetisations may be substantial. Such secondary components will be directed mainly along the Earth's field of the last few thousand years, i.e. the axial geocentric dipole field at the collecting site, although some domains acquire the direction of the field in the laboratory during storage. Domains which, for various reasons, have become blocked at different times during the last 100 million years will give only a minor contribution to the observed remanence, as the numerous reversals in the Earth's field direction during this period (Section 8.6) place approximately similar numbers of these domains in antiparallel directions.

Exposed rock outcrops may be magnetically affected by lightning. About 90% of all lightning carries a negative charge to earth, mostly discharged in 40–50 microseconds. The temperature of the plasma in a lightning strike is high, 15–20,000°C, but although this can cause intense local heating, accompanied by an explosive production of steam, any thermal remanence is also localised and is negligible compared with the magnetic effect of the current discharge. Although half of all strikes discharge less than 25 kA, it is often several hundred kA. The electrical energy is carried in a wide variety of frequencies, mostly in the 10–12 kHz range, but these frequencies are variations of a unidirectional current so the rocks only respond to the magnetisation associated with the peak direct current. The main current usually travels horizontally from the strike, passing through the higher conductivity soil and surface water, but decays exponentially with depth so that the magnetic effects are restricted mainly to the top 20 metres. Rocks

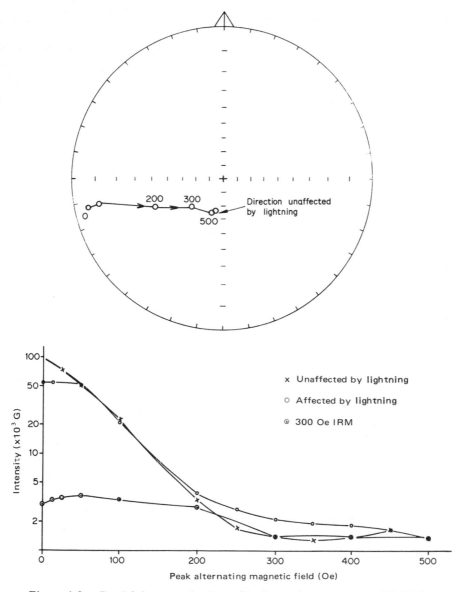

Figure 4.2   Partial demagnetisation of isothermal remanence and lightning
Alternating magnetic field demagnetisation of rock specimens affected by lightning and those given an isothermal remanence in 300 Oe show almost identical behaviour. The direction of remanence of the rock affected by lightning changes towards that of rocks which were unaffected. (After Cox 1961)

along the path of maximum current may be magnetically saturated, but at greater distances from the strike they acquire an isothermal remanence which is superimposed on their pre-existing remanence. (A small anhysteritic magnetisation may be produced if the ground and lightning currents oscillate as the lightning discharge dissipates, but such remanence, if present, will be small.) This means that most of the magnetic effects of lightning are carried by low coercivity domains and can be removed by partial demagnetisation (Figure 4.2 and Sections 2.5 and 4.5). Nonetheless, it is generally preferable when sampling to avoid areas affected by lightning such as exposed projecting points, cliff tops, etc., and rocks which deflect a compass needle. The effect of any lightning during the formation of a rock, before its burial, is lost. Rocks forming subaqueously are shielded and lightning only affects the domains of short relaxation time in cold rocks so that such remanence decays before the rock is sampled; a strong high temperature remanence could be induced in a cooling rock, but subaerial lavas cool over periods of months or years so that the relaxation times are all short and such remanence would be quickly replaced by high temperature viscous remanence in the normal ambient field.

Mechanical weathering produces little effect on the magnetic properties of the rock; thermal cycling (daily or seasonal) is small in amplitude and only effects the top few millimeters. Chemical weathering, however, can be extremely serious. Most magnetic minerals become oxidised, and finally produce iron hydroxides, such as goethite and limonite (Section 3.3). In high latitudes, this weathering of igneous rocks is mainly restricted to the surface layer and cracks, but its action in porous rocks and at lower latitudes is more widespread and may penetrate for tens or hundreds of meters causing extensive solution and recrystallisation. It is obviously preferable that weathered rocks are avoided during collecting but this can be difficult in low latitudes or in old rocks which may have been subjected to weathering on more than one occasion.

4.5   The stability of remanence

Almost all palaeomagnetic studies are concerned with the direction and intensity of the geomagnetic field at a specific time in the geological past. As secondary magnetisation is acquired during a span of time, it is generally only the primary component of remanence which is used. Since this component must be carried by domains with relaxation times greater than the age of the rock it is more stable to demagnetisation than most secondary magnetisations because of the relationship between long relaxation time (Section 2.4) and high blocking temperatures and coercivities. However, stable remanence can be acquired secondarily, particularly by crystallographic changes associated with exsolution and mineralogical disintegration as well as metamorphic events. The problem of determining the age of stable remanence will be considered separately (Section 4.6) although such age tests are in themselves a test of stability.

The simplest test is to compare the remanence of individual rock specimens before and after *storage* in the laboratory. Changes over a few months show that a large low coercivity component has been acquired and the natural remanence is completely dominated by unstable components. Apart from indicating gross instability, this test is not meaningful, as remanence which remains unchanged for several months may still be unstable over geological time, or a stable remanence may be present although obscured by larger unstable components.

The most fundamental stability test is the *consistency* of directions between rocks of similar age from the same region, and is most meaningful where different minerals are involved, such as those in igneous and sedimentary rocks. Rocks specimens will have been in different magnetic fields and subjected to different treatments since their collection so that any consistency in their directions indicates a common stable component which must have been acquired before collection. If the specimen directions show reversed magnetisation attributable to a reversed geomagnetic field, then stability is indicated since the last significant reversal of the Earth's field (Section 8.4). Even longer term stability is suggested if the directions observed are consistently divergent from the present Earth's field or the average field over the last 5 million years or so (the axial geocentric dipole field), as older geomagnetic fields only exceptionally coincided with the present pattern relative to the drifting continents.

The natural magnetisation in most rocks contains both stable and unstable components and their ratio will vary from one sample to another reflecting differences in the composition and dimensions of the magnetic mineral grains. This variation results in a *smeared distribution* in which the individual

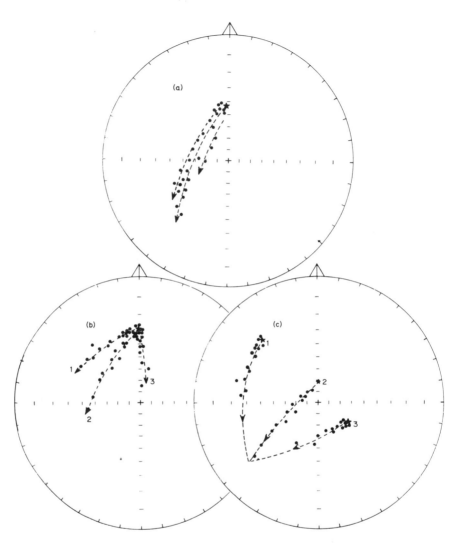

Figure 4.3 Circles of remagnetisation

(a) Rocks in the same site, or different sites, often show varying degrees of instability so that the directions are strung between the present Earth's field direction (shown starred) and the stable direction. Unless sufficient samples are stable, or sufficient stability can be isolated by partial demagnetisation, it can be difficult to define the stable direction. (b) When the measured directions are from different sites (1,2,3) that have been tilted by different amount, the directions can be corrected for tilt (c) and, if the stable direction older than the age of tilting, the original pretilt direction can be precisely defined by extrapolation of the strung distributions.

sample directions are strung between the stable and less stable directions. As the less stable components normally reflect the geomagnetic field direction of the last few thousand years, which is close to the axial geocentric dipole field, the stable direction can be estimated to be at, or beyond, the other end of the distribution (Figure 4.3). (This method of estimating the stable component is frequently used by Russian authors who describe these strung distributions as *circles of remagnetisation*). If the rocks have been unevenly tilted but not rotated horizontally, then the strung distributions form separate arcs which intersect at the stable direction after correction for bedding (Figure 4.3) and therefore allow a more precise determination.

The most effective way of isolating the stable component is by the preferential removal of unstable components by *partial demagnetisation*, particularly by thermal or alternating magnetic field methods. Increased agitation at an atomic level reduces the total relaxation time spectrum so that the domains of short relaxation time relax into their magnetically 'easy' crystallographic axes in the absence of any aligning field. Further increase in the agitation allows previously longer relaxation time domains to relax so that, if the specimen is isotropic, more domains take up random orientations and the observed magnetic moment of the specimen is increasingly composed of the longest relaxation time magnetisation (Section 2.4). Continuous measurements of remanence can be made during heating, but not during alternating magnetic field demagnetisation. However, continuous demagnetisation is slow (Section 5.5) and generally the agitation, thermal or alternating field, is increased in steps, the remanence being measured under normal temperature and field conditions after each step.

In the majority of rocks, most of the viscous remanence acquired during the last million years or so is removed by heating to 100°C, or subjecting to alternating fields of 100 oersteds (Figure 4.4). This generally causes a drop in intensity, but if the stable component is opposed to the viscous component, the intensity may increase slightly. The direction changes towards the stable direction. In some cases, higher temperatures and fields may be required as secondary magnetisations often contain higher stability components and the remanent direction changes at each step until the stable direction has been attained. In some rocks the stable component may be extremely small, less than the instrumental error of the demagnetising technique (there is a danger of adding anhysteritic, chemical or thermal components, see Section 5.5),

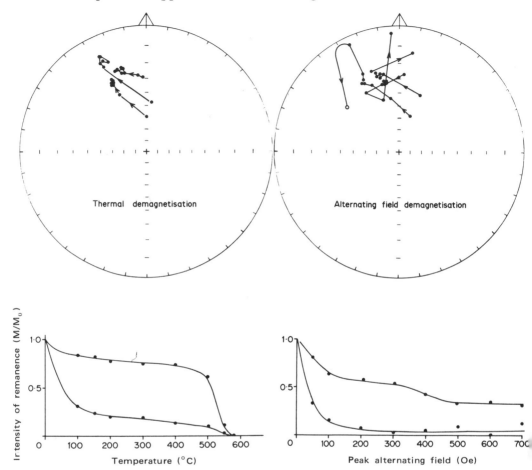

Figure 4.4   Examples of directional changes during partial demagnetisation
A variety of behaviour can occur between that for completely unstable material, which takes up random directions at successive steps of treatment, to that of highly stable material which remains approximately constant in direction. Most rocks lie between these two extremes, showing initial instability, then a stable range where the direction remains similar but the intensity of remanence continues to decrease until the remanent vector scatters as instrumental 'noise' or random magnetic moments become dominant. The demagnetisation of intensity graphs shows the behaviour of stable remanence, upper curve, and unstable remanence, lower curve.

and it may be necessary to estimate the stable direction from the strung distribution of directions produced as lower stability components are removed. However, a stable direction is frequently isolated after treatment at 150 oersteds in igneous rocks, or 300°C in sediments; this then remains constant in direction but decreases in intensity at higher treatments. Unfortunately it is impracticable to carry out demagnetisation analysis of every specimen in an extensive collection so it is necessary to select pilot specimens which can be assumed to be typical of the collection. The treatment level at which the stable direction of the pilot specimens is best isolated can then be applied to the remainder of the specimens.

The main difficulty with these demagnetisation techniques (apart from instrumental considerations discussed in Section 5.5) is the necessity to examine the complete relaxation spectrum of each pilot specimen. This is possible for many rock types by heating up to 700°C, but not for those containing minerals which are liable to physico-chemical change over this temperature range. Alternating field techniques do not introduce chemical effects but while many instruments are capable of exceeding the maximum coercivity of remanence in igneous rocks, 2000 Oe, most apparatuses are not capable of analysing the higher haematite coercivities in sedimentary rocks, in excess of 15000 Oe (Section 3.2). In general, thermal demagnetisation is most effective for predominantly haematite bearing rocks, such as red sediments, and alternating field demagnetisation is preferable for other types of rock.

Another form of demagnetisation employing *steady magnetic fields* is widely used in the U.S.S.R. A steady field is applied, exactly opposite to the direction of remanence in a specimen, after which the remaining remanence is remeasured. A higher field is then applied opposing the remaining direction of remanence, and so on. The steady field produces an isothermal remanence associated with domains of low relaxation time and a stable direction can thus be distinguished when further increases in the opposing applied field do not produce changes in direction of remanence. This system has not been widely used elsewhere as the natural remanence is carried by 5% or less of the total magnetic minerals present, and since all of the minerals respond to the steady field, the natural remanent vector is quickly swamped by the induced magnetisation.

Similar considerations apply to using determinations of the *coercivity* of isothermal remanence, saturation field, saturation moment, etc. (Figure 2.5) as these also apply to the total magnetic properties of the rock and not necessarily to the small percentage of domains which carry the natural remanence. Such properties can only be used as a coarse indication of the stability of remanence.

It may be possible to remove the magnetisation of haematite by cooling it to *low temperatures.* This is particularly valuable as haematite is often the carrier of very stable secondary remanence, produced for example by exsolution within an igneous rock or by the oxidation or reduction of other minerals. When a haematite crystal is cooled below $-15°C$ the 'easy' axes of magnetisation change direction from within the basal plane to become parallel with the ternary axis (the Morin transition). On warming back to room temperature, the magnetic direction in each crystal returns to the basal plane and, in the absence of an aligning force, would be randomly oriented in isotropic rocks so that the remanence associated with haematite would be lost. Unfortunately this low temperature behaviour is complicated by small grain sizes and is affected by the perfection of the crystal structure and the purity of the composition so that, in many rocks, the temperature of the Morin transition is lowered. On reheating, some crystals are able to remember their previous alignment so that some of the original remanence is regained (Figure 4.5). A similar transition is observed in magnetite at $-143°C$ but, in addition, magnetite changes its crystallographic form from cubic to orthorhombic when cooled below $-154°C$, with a corresponding change in properties. At present, the most useful application of these low temperature methods is in the identification of minerals rather than the testing of stability.

In sedimentary rocks the remanence may be carried by the cement or the detrital grains (Section 4.2) and it is sometimes possible to isolate the remanence of the detrital grains by preferentially removing the cement during soaking in *acid.* The specimen can be soaked for varying periods so that the change of remanence can be studied as the cement is progressively removed. The acid penetration is assisted if thin disks are used, although the remaining magnetic moment may be difficult to measure. It is quicker if the acid is forced through under pressure, and this also tends to concentrate the acid's action inside the porous channels. Some successful results have been obtained although the main object has usually been to determine the carrier

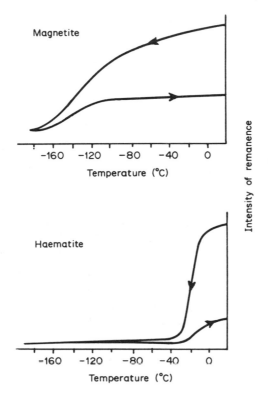

Figure 4.5   Low temperature behaviour
(a) Magnetite shows transitions at low temperatures, between -140 and $-160°$C. (b) Natural haematite shows a transition at approximately $-15°$C, but has an ability to remember the previous direction when it is warmed again. (The decrease in remanence as the transitions are approached can give the appearance of partial self-reversal (Section 7.2).)

of remanence and the technique has not yet been used on a large scale. The main problem is to prevent the solution of detrital grains before the cement has been removed and so it cannot be readily applied to impervious rocks or to those with silicate cements. This test, of course, only isolates the remanence associated with detrital grains; further tests must then be applied to establish its stability and age.

4.6   The age of remanence

If a stable direction can be isolated it can be compared with other observations (palaeomagnetic, palaeoclimatic, geological, etc.) and if consistent with these, can be presumed to be primary in origin. However, there are objective tests which indicate if the stable component is primary i.e. of the same age as the rock itself. The most fundamental of these is the fold or tilt test.

As a primary magnetisation is acquired in a uniform field it will be of uniform direction throughout any one rock formation. If samples are collected from an area where subsequent geological events have either folded or broken the formation into blocks which have tilted in different directions obviously the primary remanence will also have been similarly tilted (Figure 4.6). The observed directions of each sample must be corrected by the angle of tilt which can be measured from the orientation of the rock strata (section 5.3) and after these corrections primary directions will be uniform and distinguishable from any directions of magnetisation acquired after tilting. The tilt correction may not be simple, as complex rotations may also be involved. There is still a possibility that the uniform magnetisation is not primary but as it has remained stable during and subsequent to the tilting it must at least have been acquired prior to the folding or tilting of the rocks and, in view of the exponential nature of relaxation time, most probably at the time of the rock's formation.

The age of remanence can also be assessed by studying magnetisation in conglomerates (Figure 4.6b). These are formed by the very rapid turbulent deposition of eroded material from flash floods, turbidity currents, etc., and contain large pebbles or boulders which are usually randomly oriented due to their rapid deposition. Similarly slumped material will become randomly oriented. If the magnetisation of these pebbles is uniform, then it must have been acquired subsequent to its deposition, but randomly scattered directions imply a very long term stability. If the conglomerate or slump is very old, then the rocks from which the pebbles were derived must also be stable for similar periods of time and therefore such remanence is probably primary. The main difficulties with this test is that it must be possible to identify the source of the rocks in the conglomerate and many different pebbles must be examined to make a statistically satisfying test of either uniformity or random scatter of directions (Section 6.3).

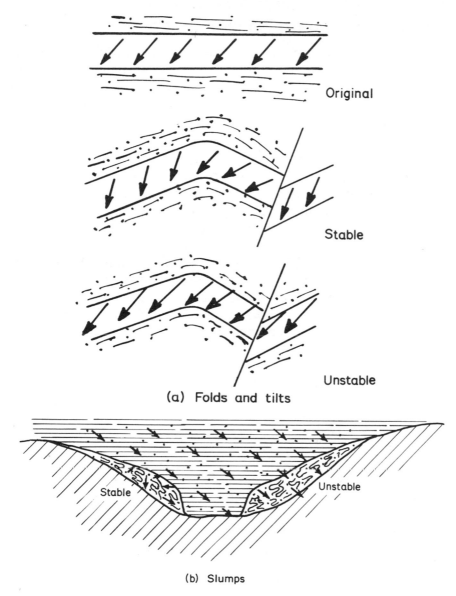

Original

Stable

Unstable

(a) Folds and tilts

Stable

Unstable

(b) Slumps

Figure 4.6   The tilt and slump test for stability
Rocks which have been (a) folded, tilted or (b) slumped since their formation will show uniform directions in their present position if their remanence post-dates the movement, but will be uniform, after correction for the movement, if the remanence predates the tectonic disturbance.

49

In an igneous rock, the probability that the stable component is primary can be assessed by comparing the intensity of natural remanence remaining after heating through various temperature ranges with that acquired over the same range of temperatures when cooling from above its Curie temperature in a known magnetic field. If there are no physico-chemical changes during these heating experiments, then a constant ratio between the natural and thermal remanent intensities suggests that the natural remanence is of thermal origin over that temperature range where the ratio is constant and is therefore primary; differences in the ratio usually appear at lower temperatures when secondary remanence and decay of primary remanence are most common. Similarly comparisons can be made between the natural remanent intensity and the amount of viscous remanence the rock can be expected to have acquired. This can be predicted by holding the specimen at set temperatures in a known field for various periods of time, up to several days. From the growth rate of viscous remanence during these periods, an estimate can be made of the maximum amount of viscous remanence which would be acquired in various circumstances (Figure 4.7). For example, if a

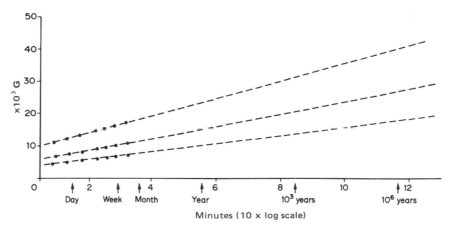

Figure 4.7   The acquisition of viscous remanence
The rate at which the remanence is acquired depends on the temperature and field at which it is held. At each temperature the growth of viscous remanence is logarithmic and the amount of viscous remanence which would be acquired if held at any one temperature for prolonged periods can be estimated by extrapolation from laboratory observed values. In this example, a basalt sample was held in a field of 2 Oe at temperature of 20°C, 100°C and 300°C for various periods of time.

rock is buried beneath some 4 km of younger rocks it will acquire a viscous remanence at a temperature of $100°C$ while buried. However, neither of these comparative intensity tests can be carried out on a large number of specimens and such detailed experiments are more often designed to determine the ancient intensity of the Earth's field (Section 7.4) or to investigate the geological history of the samples (Section 9.5).

The age of the rocks, and therefore of any primary remanence, is determined by either standard stratigraphical techniques or radioactive methods. Some radioactive methods can also indicate geological events subsequent to the rock's formation which may have introduced a stable, secondary component. Radioactive dating relies on the retention of daughter isotopes, or fission tracks, which are produced at a known rate from the radioactive decay of the parent isotope. It is therefore essential that the minerals are able to retain the daughter products, such as the A^{40} produced by the decay of K^{39}. Since the argon retentivity of different minerals varies, a moderate thermal event may drive the argon from some minerals, but not others. As this thermal event may also result in the growth of a higher temperature viscous remanence and probably encourage exsolution processes, it may be accompanied by the production of a stable, mainly chemical, secondary magnetisation. Rocks containing minerals with different K^{40}/A^{39} ratios may therefore be suspected of having a stable secondary component. In general, it appears that the total argon retentivition of a rock is more affected by thermal events than the primary remanence of their magnetic minerals so that constant isotopic ratios for different minerals within a rock implies that any stable remanence is primary.

It is possible that further developments in radioactive dating techniques may also be very significant in palaeomagnetic work as it may soon be possible to date the magnetic minerals themselves. Uranium can enter the crystal lattices of either titanomagnetite or haematite, so that the age of the individual magnetic crystal lattices in igneous, metamorphic or sedimentary rocks could be dated from fission tracks or the daughter products of uranium radioactivity. This means that the age of haematite could be determined within a sediment, or the time of exsolution of a titanomagnetite in an igneous rock. If these ages are the same as the age of the rock itself then stable remanence associated with such crystals would almost certainly be primary, but if, for example, exsolution took place much later then its component minerals could only carry a secondary remanence.

4.7   Magnetic anisotropy and inhomogeneity

Magnetic anisotropy arises from a net alignment of either the shape or crystallographic axes of magnetic minerals within a rock (Section 2.5). Most basic igneous rocks are isotropic, but anisotropy can arise by alignments caused by either flow within the liquid magma or the gravitational settling of minerals as they crystallise. Flow banding is more common in acidic rocks, particularly rhyolite, and gravitational settling occurs in several massive basic and ultrabasic bodies. Both processes of alignment take place while the magma is still fluid, that is at temperatures well in excess of the Curie temperature, but minerals forming below the Curie temperature, for example, by exsolution, will also have a crystallographic and shape alignment directly related to the orientation of the host minerals, for they develop along preferred crystallographic planes within them. In metamorphic rocks minerals may be oriented by direct pressure, but their anisotropy is usually due to plastic flow and crystal growth in alignments perpendicular to uniaxial pressures (Sections 2.5 and 4.3). Sediments normally show distinct alignment of minerals in their bedding planes, but this alignment is mainly of platey minerals, particularly micas, which are non-magnetic and the primary remanence is carried by minerals which have been magnetically aligned after deposition and do not therefore reflect alignments at the time of deposition.

The magnetocrystalline anisotropy in most magnetic minerals is lost at temperatures above $300°C$ and it is the anisotropy due to shape which is of greater importance in palaeomagnetic studies. As this is mainly associated with larger particles, most magnetic anisotropy is associated with material of low relaxation time and the more stable remanence, carried by smaller particles, may be isotropic. This also means that the remanence associated with more anisotropic grains is preferentially removed by partial demagnetisation tests.

Obviously anisotropy must be detected before attempting to define the ancient field as anisotropic specimens have a direction of primary remanence which is a compromise between the ambient field direction and the net alignment of the magnetocrystalline axes or the shape of the particles. The degree of anisotropy of remanence in igneous rocks can readily be tested by heating and cooling the specimen in a field of known strength and direction; if the thermal remanence it acquires departs significantly from the applied field, then its primary remanent direction will also have been affected by the anisotropy of its minerals. Other methods of measuring anisotropy (Section 5.4) are difficult to interpret as they determine the anisotropy of all magnetic material present and not just the anisotropy of those grains which carry the primary remanence. It is generally safer to exclude anisotropic specimens from analyses defining the ancient geomagnetic field. Such measurements in metamorphic rocks, however, can be extremely useful in determining the direction of tectonic pressures, etc. (Section 9.3).

As the remanence of most rocks is carried by only a small percentage of the magnetic material present, it is possible that the carriers of remanence are not evenly distributed throughout the rock. This occurs particularly in sediments where the magnetic particles tend to occur in layers. This inhomogeneity of distribution within a specimen can usually be detected during measurement (Section 5.4) when it renders accurate determination of the true direction of remanence more difficult. It is not really practical to correct for such effects and it is simpler to eliminate these specimens from any palaeomagnetic analysis. If, however, inhomogeneity is not detected during the measurement of individual specimens, its effect would be random from one specimen to the next and would result in a wide scatter of different specimen direction. Such a scatter would probably be taken to indicate instability, and the specimens would again be eliminated.

4.8   Summary and comments

Although the processes involved in the magnetisation of rocks are extremely complex, there is only rarely any serious error in assuming a simple state of affairs in which igneous rocks acquire a thermal remanence at the time of formation and sedimentary rocks acquire a remanence at or shortly after the time of deposition. Superimposed on these primary magnetisations are viscous magnetisations which are acquired over prolonged periods of time. The direction and intensity of primary remanence in a rock can be isolated by a variety of laboratory and field tests, some of which are capable of indicating its age. The main problems are in isolating any stable remanence and in determining its age and relationship to the ancient field.

The main complexities arise if the rocks have undergone physico-chemical changes since their formation; such effects are most likely if the rocks have been subjected to deep burial or tectonic forces. Under such circumstances the simple assumptions cease to be valid and considerable caution is required before deductions are made concerning the previous nature of the Earth's magnetic field.

5

Sampling and measurements

5.1 Introduction

It is not possible to measure the various components of magnetisation of rocks in the field, so oriented samples must be collected at selected points and the samples measured and analysed in the laboratory. The methods of collection and orientation are described, followed by an outline of the physical basis of standard palaeomagnetic instruments and of the problems which arise in their practical use. Technical specifications are not given as these are variable and are detailed elsewhere (see bibliography). Finally an assessment is made of the accuracy of these instruments and the accuracy of collection and orientation of samples in the field.

5.2 Sampling rock formations

With the exception of deep-sea sediments, rocks are generally formed intermittently and, as a result of their subsequent development and erosion, their present day outcrops are irregularly distributed. Consequently a continuous record of past changes of the Earth's magnetic field can only be obtained from deep-sea core samples and, for most of geological time, only a fragmentary record is available from which the history of the geomagnetic field must be compiled. Any formation must be sampled at a series of discrete locations, *sites*, where the rocks can be assumed to have acquired their primary magnetisation simultaneously, and the sites must, as far as possible, be uniformly distributed through the time span represented by the rock formation. In igneous rocks, a site can be readily defined as a single small lava flow dyke since these cool and became magnetised at a specific time. A site in sedimentary rocks is more difficult to define, unless the process of its magnetisation is known, and so a site is generally considered to consist of rocks distributed along a few metres of a single stratigraphical horizon. Several different samples are required from each site in order that instrumental and collection errors can be averaged and the statistical validity of the observations can be assessed. The number of samples depends on the statistical reliability required, the nature of the rock and also on the difficulties involved and time available for collection; in general, a minimum of six separately oriented samples is sufficient for reliable statistical analysis (Chapter 6), but two or three separately oriented samples per site may be adequate for a reconnaissance survey. The samples may be hand samples, about 15 x 7 x 7 cm, from which cores, generally 2.5 cm diameter, are subsequently drilled, in the laboratory, or separately oriented core samples, 10–15 cm long, may be drilled in the field (Figure 5.1). Specimen cylinders or disks, usually 1.0 to 2.5 cm high, are cut from the cores for measurement of their magnetic properties (Section 5.4).

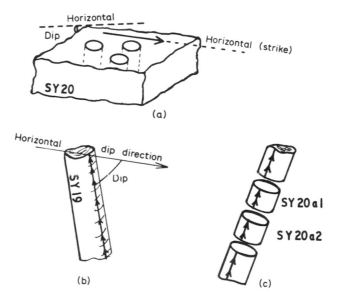

Figure 5.1   Samples and specimens
(a) Hand samples are oriented on one surface and cores drilled from them later in the laboratory. (b) Sample cores are drilled and oriented in the field. (c) The cores are subsequently sliced to provide suitable sized specimens for measurement.

Generally only one specimen need be cut per core. Variation in the magnetic properties of specimens cut from the same core are caused by random instrumental errors, inhomogeneous magnetisation or differences in the age of magnetisation within the core. The magnitude of instrumental errors is usually known from repeatability experiments (Section 5.4) and inhomogeneity or significant age differences can be distinguished by comparison of measurements between different samples so that only rarely is the extra information gained from more than one specimen per core justified by the large increase in the number of measurements. The remaining material from the core is, of course, available for Curie temperature analysis, thin sections, radioactive dating, etc. Three or more cores are, however, required from hand samples as it is necessary to average out errors in drilling and in transferring orientation marks (Section 5.3) from the surface of the hand sample to the cores.

Drilling of cores in the field is more expensive, initially, than the collection of hand samples, but the additional cost is more than offset by the higher accuracy and the advantage that fewer specimens need to be measured. For reconnaissance purposes, or where the rocks are excessively hard, hand sampling is suitable, but detailed surveys require large numbers of observations for which it is impractical to collect and measure a sufficient number of hand samples. Field drilling also has the advantage that it is much more versatile, so that sampling is not restricted to fractured or projecting rocks and fresher material can be collected. As the freshness of the rock is of fundamental importance (Section 4.4), extra minutes spent locating the freshest, unweathered material and that least likely to have been affected by lightning, not only saves hours of later laboratory work, but also increases the probability of obtaining meaningful results. Samples from stream sections, road cuttings, quarries, etc., are generally less weathered and where the weathering is deep deeper drilling methods (5–10 m) or explosives can be used to penetrate the weathered layer.

5.3   Orientation of samples

The surface of a hand sample or the side of a field core is marked with a fiducial mark, usually either horizontally or down the maximum slope, and the orientation of this mark must be determined before removing the sample. Subsequent measurements can then be related to this known

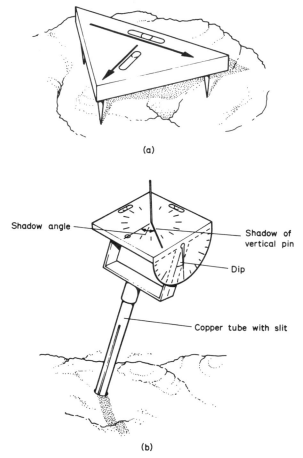

(a)

Shadow angle

Shadow of
vertical pin

Dip

Copper tube with slit

(b)

Figure 5.2   Instruments for the orientation of rock samples
A flat surface for the orientation of rocks can be obtained simply by using (a) a tripod for
hand samples and (b) an orientation table for field cores. The dip and attitude of this
surface can then be determined and a fiducial mark is then obtained using either a scriber
along a slit in the orientation tube or by marking the position of the legs of the tripod.

orientation mark which is usually defined (Figures 5.1 and 5.2) in terms of
its *dip* (the maximum angle of slope of the surface; positive below horizontal
and negative above) and the direction relative to geographic north of either
the maximum slope or the horizontal line perpendicular to it (*strike* line).
The amount of dip can be measured relative to either a plumb line or

levelling bubbles (Figure 5.2). The direction can be determined by several methods. The *magnetic compass* is fast, cheap and versatile and has been widely used despite the possibility of systematic error, particularly in igneous terrain, where distortions of the local geomagnetic field are most common. The magnetic compass can also be used to obtain bearings of the fiducial mark relative to distant *topographic* features, or alidades can be constructed for this purpose. The direction of these bearings can then be obtained from topographic maps and the direction calculated relative to geographic north. The highest precision is obtainable using a *gyrocompass*, but these are expensive and, although unaffected by visibility and magnetic disturbances, they are difficult to use in windy conditions. They are also much more precise than the accuracy with which the fiducial line can be marked on specimens ($1-2°$). The fastest, most accurate, lightest equipment is the *sun compass* which, at its simplest, consists of a horizontal platform (required in any case for determining the amount of dip) with a central, vertical pin. The angle of the shadow of the pin is noted relative to the fiducial mark, and the time, date and location noted for subsequent calculation of the orientation (Figure 5.2). (The azimuth of the shadow of the pin is then given by $180 + \tan^{-1}$ ($-\sin$ LHA/ ($\cos \lambda \ \tan$ SD $- \sin \lambda \ \cos$ LHA)) where λ is the latitude, LHA is the local hour angle and SD is the sun's declination. SD and LHA are available in Air and Nautical Almanacs in which LHA is given by the Greenwich hour angle + 0.25 x longitude.) The sun compass is least accurate in low latitudes at midday, i.e. when the length of the shadow is less than one tenth of the length of the pin, but is otherwise very fast and precise, being accurate within $1°$ if the time is known within two minutes and the location within $0.5°$ except with one hour of midday at low latitudes when, in any case, the shadow is too short to measure easily. Comparison of the four directional methods in igneous terrain shows that the gyrocompass, topographic and sun compass agree with each other within $1½°$, but magnetic orientations differ by $30°$ or more. The method used clearly depends on local conditions, but the magnetic method should only be used in igneous terrain as a useful adjunct to indicate the possibility of lightning effects, which distort the local geomagnetic field. The total accuracy of the orientation should therefore be within some $2°$, which is comparable with the accuracy with which the orientation marks are scribed on drill cores. However, this accuracy is reduced when using hand samples as further errors are involved during re-orienting the samples for drilling and during the transference of orientation marks to the individual cores. The accuracy of such orientation probably drops to 3–5° and may be partly systematic if caused by incorrect re-orientation.

The orientation of the sample allows its magnetic properties to be related to its present day position, but many rocks have been tilted since their formation. It is necessary to correct the directions of primary remanence for this tilt by use of measurement of the dip and strike of the present orientation of the original horizontal surface. This surface can be defined as the bedding plane in fine-grained sediments, mudstones and shales, which, on deposition, have shallow, wet angles of rest and is unlikely to have lain consistently at angles greater than 2–3° to the horizontal at the time of deposition; coarser grained sediments (particularly aeolian) and igneous rocks may have steep angles at their formation and are clearly unsuitable for determining the original horizontal. In fact, the largest errors in palaeomagnetic measurements occur in the determination of bedding tilt. This is mainly because the finer grained deposits are more easily distorted (less competent) and, as they are buried by younger rocks, differential compaction of alternating competent and incompetent beds results in substantial variation in their tilt. These variations can be largely averaged by eye and the best average bedding plane determined, but variations of 5–10° still occur in the bedding estimates at different sites within a uniformly tilted sequence. If the tilting is not uniform, e.g. in folded beds, then correction may also be necessary for the plunge of fold axes as this causes a rotation of the rocks which is not defined in their present strike and dip. Similarly, a uniformly tilted sequence may have been rotated about a horizontal axis, but it is generally impossible to determine this rotation from the geological evidence and it is more common for the palaeomagnetic method to be used to determine such rotations (Section 9.3).

5.4 Measurement of remanence, susceptibility, anisotropy and inhomogeneity

A variety of instruments are available for measuring the intensity and direction of remanence of rock specimens, and similar instruments are modified for the measurement of archaeological or extraterrestrial material.

Two basic types of instrument are used for routine measurements, the astatic and the spinner magnetometer; the use of other types, resonance, ballistic, etc., are generally restricted to specialised experiments.

(a)   Astatic Magnetometer.

This consists of a magnet system, suspended on a torsion fibre, in which the orientation of the magnet is arranged astatically so that the system responds to the magnetic field of the specimen but is insensitive to geomagnetic and laboratory magnetic disturbances (Figure 5.3). The choice of torsion fibre restricts the sensitivity of the instrument so that, in general, one instrument covers a specific range of magnetic moments. The sensitivity of a single instrument can also be varied by use of an external magnet, avoiding changes of the suspension fibre. The deflection of the magnet system by the field of the specimen in different positions allows the magnetic vector in the specimen to be determined. A uniformly magnetised specimen has a field that is similar to a dipole magnet at its centre. By assuming a specimen is

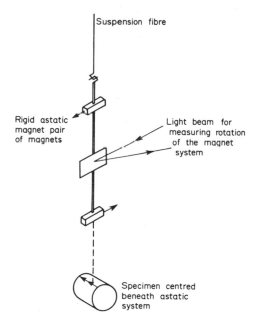

Figure 5.3   **An astatic magnetometer**
The deflection of the suspended magnet system is proportional to the direction and intensity of magnetisation of the specimen beneath it.

uniformly magnetised, the number of observations can be reduced. This assumption is usually checked by taking more than the minimum number of readings to define a dipole so that any disagreement between different readings can be examined in terms of either *inhomogeneity* or *anisotropy*. The major difficulty with this instrument is its sensitivity to vibrational disturbances. This can be reduced by vibrationless mounting, selection of the torsion fibre so that the response time differs from that of the dominant source of vibration, etc., but in general, this means that very quiet conditions are required for measurements of weakly magnetic rocks (intensities less than 10^{-5} G cm^3). On short period, less sensitive instruments, some $10-12$ specimens can be measured per hour and on more sensitive instruments $6-10$ per hour. These rates can be increased to $30-40$ and $15-20$ per hour respectively if the specimen is rotated near the magnetometer at a faster rate than the response of the magnet system so that the magnetometer then responds to the average magnetic moment along the axis of rotation. This means that inhomogeneity has less effect and the minimum number of observations can be used although care must be taken that the magnetisation of the specimen still lies within the accurate range of measurement of the instrument as no repeat measurements are usually involved for checking the precision of the readings.

(b)   Spinner Magnetometer.
In this instrument, the specimen is spun within or near to pick-up coils in which it induces a alternating voltage, whose amplitude depends on the intensity of the component of magnetisation along the axis of spin and perpendicular to the axes of the coils; the phase of the direction of this component is measured relative to a fiducial mark on the rotating head (Figure 5.4). Early versions of this instrument were inaccurate for weakly magnetised specimens, but improved electronic techniques, particularly in the measurement of the phase, mean that these instruments can now cover the range of intensities found in most rocks with equal precision, although weakly magnetic rocks must be rotated either faster or for a longer time to allow the signal to be integrated above the noise level of the instrument. The noise is of two types, mechanical and electronic. Mechanical noise arises from the motion of the rotating head and is reduced if the rotation is by air turbine rather than direct motor drive, but this advantage is offset by an

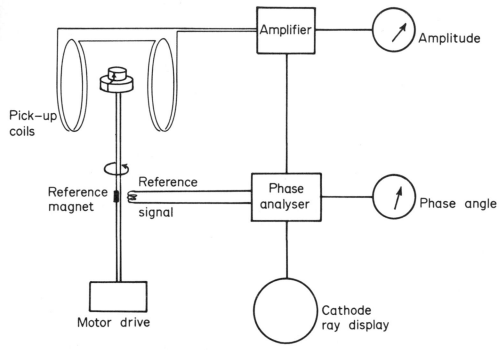

Figure 5.4   A spinner magnetometer
The rotation of a rock specimen within the pick-up coil induces a voltage in the coil, the phase of which depends on the direction of remanence, and the amplitude on the intensity.

increased electrostatic charge on the rotating parts. (The electrostatic disturbance is particularly significant in measurements of weak rocks, and can be reduced by using extremely fine water jets to dissipate the charge.) The major source of noise is electronic which is largely overcome by integrating a weak signal over time intervals of a few minutes. Generally some 10–15 samples can be measured per hour with this instrument, but this drops to 5–6 per hour for weakly magnetised specimens. One advantage of this method of measurement is that the rotation tends to average out inhomogeneity effects. This means that the minimum number of observations can be used to define the specimen's magnetisation and furthermore the assumption of uniform magnetisation can be checked as the output signal can be displayed and compared visually with the sine wave signal expected for a uniformly magnetised specimen.

(c)   Fluxgate magnetometers.

The advantages of the astatic and spinner magnetometer systems are combined in sensitive fluxgate and second harmonic detection devices. These can be arranged in astatic pairs (or more complex arrangments to reduce the effect of extraneous magnetic noise) but, as they are not suspended, they are comparatively insensitive to vibrational noise and can either be positioned to measure simultaneously the magnetic field of the specimen along different axes, or the specimen can be rotated near the fluxgates. It is probable that these devices will be further improved in the near future, and portable magnetometers using fluxgates have already been made for measuring the remanence of specimens in the field. However, it is not possible to carry out full palaeomagnetic analyses without the use of other instruments, particularly for demagnetisation, and so later laboratory work is still essential.

(d)   Low Field Susceptibility.

Low field susceptibility can be determined using astatic or spinner magnetometers to measure the remanence induced in a specimen by a weak field, such as the Earth's. A magnetisation of the specimen is measured in opposite directions so that the natural remanence is reversed, but not the induced remanence, so that the amount of induced remanence, and thus the susceptibility, can be calculated. Low field susceptibility can also be determined using the transformer *susceptibility bridge* (Figure 5.5). This consists of two ferrite rings with primary winding carrying an alternating current which produces an alternating magnetic field of up to 10 oersteds across gaps cut in the rings. A ferrite plug can be positioned to balance the two circuits so that a specimen placed within one of the gaps unbalances the circuit in secondary windings in proportion to its susceptibility. This instrument is fast and can be designed for the rotation of specimens within the gap, allowing the isotropy of susceptibility perpendicular to the axis of rotation to be determined. The uniformity of susceptibility is a measure of inhomogeneity and anisotropy, but is difficult to measure systematically for standard specimen disks or cylinders because of their non-symmetric shape. The use of an alternating current can, of course, result in the acquisition of an anhysteritic magnetisation, but this is exceedingly small and the susceptibility measurement can, in any case, be made after the completion of measurements of natural remanence.

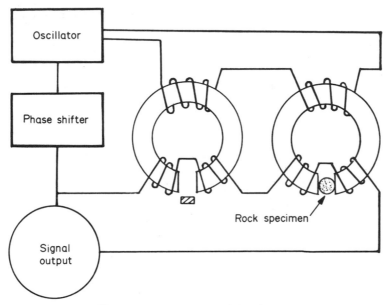

Figure 5.5   A susceptibility bridge
The rock specimen unbalances the circuit by an amount which is proportional to its susceptibility.

(e)   Anisotropy.
Anisotropy can be determined by measurements of susceptibilities in different directions in either weak or strong applied fields using standard magnetometers or susceptibility bridges, but anisotropy in strong fields is more commonly measured by suspending the specimen in the field and observing the torque exerted on the suspension fibre; if there is a preferred direction of magnetisation, the specimen attempts to turn into it, exerting a torque on its suspension fibre. The torque in different positions can therefore be used to determine the direction and magnitude of the anisotropy. In routine palaeomagnetic studies it is generally sufficient to determine the maximum and minimum values of susceptibility of a specimen – those in which the anisotropy (χ_{max}/χ_{min}) exceeds $\pm 5\%$ are generally regarded as unsuitable for use in determining previous field directions. These techniques can be applied to analyses of the magnetic fabric of the rock (Section 9.3) and it is then necessary to define the shape

of the anisotropy ellipse. At least ten separate determinations are required to define this ellipse and even then this analysis can still only be considered as a generalisation about the nature and orientation of the magnetic minerals. Nonetheless, this information can be gained rapidly and the use of visual displays, direct computer analysis or digital outputs from spinner or fluxgate magnetometers may eventually lead to precise analysis.

(f)   Inhomogeneity.
Inhomogeneity of magnetisation is mainly reflected in variations in measurements of remanence and susceptibility of a specimen at different orientations. The field of a uniformly magnetised specimen then corresponds to that of a dipole at the centre of the specimen and evenly distributed magnetic material will give constant susceptibility in all directions. The value of inhomogeneity is difficult to quantify and, if determined by susceptibility observations, may be obscured by anisotropic effects. Until the magnetisation of individual grains within a rock can be studied, which is now becoming possible, it is difficult to use materials showing inhomogeneous magnetisation and if the repeated measurements of specimen remanence show scatter above a certain amount such specimens are generally eliminated from subsequent analysis. Inhomogeneity is sometimes reduced during demagnetisation techniques, but in most cases these methods appear to have little significant effect on inhomogeneous characteristics.

5.5   Demagnetisation

Analysis of the relaxation time spectrum of individual specimens or isolation of components of long relaxation time is undertaken by subjection of specimens to two main processes, temperature changes or alternating magnetic fields. Both processes operate in a similar way (Sections 2.4 and 4.5) and the difficulties in both are the same; the removal of low relaxation time remanence without adding new, spurious magnetisation.

Thermal demagnetisation may be undertaken continuously by heating the specimen beneath a magnetometer and noting changes in remanence. The observations of remanence, particularly of weakly magnetic specimens, may take several minutes and it is necessary to maintain a steady temperature during the measurement. Therefore the demagnetisation proceeds in a series of steps, rather than strictly continuously, but there is less likelihood of introducing spurious thermoremanent magnetisation as the specimen is not cooled between each step. In practise, it is difficult to make a complete measurement of the remanence during continuous thermal demagnetisation as there is little room for placing a furnace beneath the magnetometer with access for turning and rotating the specimen. Consequently most experiments are confined to changes of remanence in two dimensions, thereby avoiding turning the specimen. The need for thermal insulation between the magnetometer and the furnace also means that the detecting heads must be distant from the specimen, and therefore only the more strongly magnetic rocks can be investigated by this technique.

Furnaces, electrically or gas heated, can be made in various sizes so that up to 100 specimens can be heated at a time; the number being controlled by the volume of field-free space which can be maintained within the furnace. These specimens can be heated to a set temperature and then cooled in zero field so that the lower relaxation time magnetisation becomes blocked in random directions and the remanence of the higher relaxation time material can then be measured with standard magnetometers. A stray field over the specimens during cooling will, of course, result in a spurious thermoremanent magnetisation. Rocks vary in their sensitivity to this field, but fields as low as 5 γ have detectable effects on the remanence of some rocks. The most serious problem in both continuous and stepwise thermal demagnetisation procedures is the possibility of chemical changes occurring during heating. Many magnetic minerals tend to oxidise during heating and some are reduced. Some of these chemical changes can be reduced by heating in an inert atmosphere or vacuum, but the presence of various gases within the rocks themselves and the chemical instability of some minerals at moderate to high temperatures means that these effects are not completely prevented. Clearly, rocks containing mostly haematite, such as red sediments, are least affected by these processes which can be monitored, on a qualitative scale, by noting changes in susceptibility of the rocks as a whole during the progressive heating.

Alternating magnetic field demagnetisation is carried out by passing an alternating current through a coil, thereby producing an alternating field along its axis (Figure 5.6). The current is usually that of the public mains supply, alternating at either 50 or 60 Hz, which is controlled by a potential

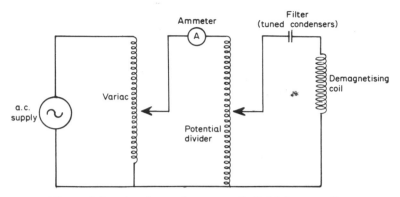

Figure 5.6   An alternating magnetic field demagnetiser
The alternating current is smoothed using tuned filters and passed through a coil so that a specimen within the coil is demagnetised by the alternating field along its axis. In order to demagnetise the rock specimen in all directions, the specimen is usually rotated within the demagnetising coil.

divider to avoid sudden steps in the current. The specimen is placed on the axis of the coil and the Earth's magnetic field cancelled (Section 5.6) to reduce anhysteritic magnetisation. The mains supply does not have a pure wave form so it is necessary to filter the supply to achieve as pure a wave form as possible; irregularities in the wave form result in a unidirectional field being produced along the axis of the coil so that the specimen can acquire an anhysteritic magnetisation. Higher frequency currents can be more effectively, but more expensively, filtered but in standard apparatus this spurious component can still be kept below 1% of the total current, and in most rocks the anhysteritic magnetisation which results is not significant in comparison with the natural remanence until high fields are used. In order to reduce the effect of any anhysteritic magnetisation and also to demagnetise the specimen as a whole, rather than along separate axes, most apparatus contains a tumbling device in which the specimen is simultaneously rotated about two or three mutually perpendicular axes while the alternating field is applied; this presents the specimen in as many different orientations as possible to both the demagnetising field and the anhysteritic component. Unfortunately the geometry of the tumbling devices prevents a completely uniform presentation of all directions in the specimen and therefore it is important that there is no correlation between the tumbling action and the frequency of the alternating field and the specimen must not be systematically rotated relative to the field but should be presented randomly to the field along the axis. In practise, the space restrictions mean that the specimen can usually only be inserted within the tumbler in one or two positions so that the rotation is generally systematic, but this effect can be eliminated to a large extent by measuring the remanence after one demagnetisation, and then reversing the orientation of the specimen within the holder and demagnetising at the same field; the average values of remanence then being taken as the true value at that demagnetisation level. The limits of this apparatus are therefore set, at the moment, by the purity of the wave form which can be produced; this is more critical at high fields. Most conventional alternating field instruments still have upper limits of 1000 to 2000 Oe, although experimental apparatus has been made for producing alternating fields of 12000 Oe. It is probable that, in the near future, improvements in filtering will allow the full relaxation spectrum to be investigated in this way for all rocks; the highest field required being of the order of 20000 Oe for the demagnetisation of fine grained haematite, common in red sediments.

At the moment, the introduction of extraneous magnetisations can be prevented most efficiently during thermal demagnetisation, but this method is particularly subject to chemical changes, particularly oxidation, of the magnetic minerals and is therefore best suited to haematite-bearing sedimentary rocks. The minerals in igneous rocks tend to be more sensitive to heating and are therefore more effectively treated by the use of alternating magnetic fields. Conveniently, the coercivity of most minerals in igneous rocks is low and within the present technical limits of the instruments. However, it is in this demagnetisation procedure that technical improvements are needed to allow for the full analysis of the relaxation spectrum of rocks containing high coercivity minerals.

5.6   Field free space

Many palaeomagnetic experiments, particularly demagnetisation, require a field free space to be maintained for various periods of time. This can be done in two ways, either by cancelling the field by an exactly opposite field or by high permeability magnetic shielding. A current passed through a loop produces a magnetic field which can be varied so that at some point its field cancels the ambient field, but the field gradients produced are strong and the field free space is generally inadequate. A more satisfactory system utilises pairs of coils (Helmholtz pairs), which may be round or square, arranged so that different components of the ambient field are cancelled by each pair. The arrangement of the coils with diameters of 2 m or more allows the field to be cancelled within some 5γ in a volume of some $100\,\text{cm}^3$. Better cancellation is generally impractical as the residual field is mainly formed by varying field gradients associated with the laboratory. Such gradients can be reduced by occasional manual monitoring of the field and correction of the current in the compensating coils, or by continuous monitoring using fluxgate detectors and the application of feedback systems.

Magnetic shielding using mu-metal shielding is extremely expensive in material and design time, particularly as the shielding must be continuous. Other than for the storage of material, or long term, self-regulating experiments, the use of mu-metal shielding is impractical except for the initial construction of rooms within which palaeomagnetic experiments can be conducted. Such magnetically shielded rooms usually require further field control within them and it is still uncertain how they react to high frequency alternating magnetic fields such as those used in demagnetisation. Under most circumstances, it is therefore cheaper to construct a laboratory in quiet magnetic and vibrational locations and to control magnetic field changes by compensating coils.

5.7   The accuracy of palaeomagnetic techniques

The intensity and direction of remanence of rock specimens can be measured with high accuracy; repeated measurements of stably magnetised specimens show a repeatability within $\pm 1°$ in direction and $\pm 3\%$ in intensity, although some instruments are only capable of this precision over specific ranges of intensity of magnetisation. The direction of remanence can therefore be measured with greater accuracy than that attainable when collecting and orienting the specimens in the field so that instrumental improvements need to be in the speed of measurement rather than in precision. Individual samples can be oriented within $1°$, although this precision is often lost within the fiducial mark itself, generally $1-2°$ wide, and more serious errors are introduced if orientation marks need to be transferred from one surface to another, as in the case of hand samples. There is a clear danger that spurious magnetisations are added when analysing the natural remanence by incremental demagnetisation. With care these magnetisations can be kept down to a level at which they have no significant effect on the natural remanent vector, until such high fields or temperatures are reached that any stable direction of remanence will have already been adequately isolated. Thus it is possible to define the direction of natural remanence associated with most parts of the relaxation spectrum of a single sample within about ± 3°.

The errors in measurement and collection are not generally systematic, and can therefore be largely averaged out when calculating the mean direction of several samples from the same site. The most serious source of error is that of determining the original horizontal, i.e. the amount of tectonic tilt that has occurred since the formation of the rock. This can be partially averaged out between different sites, but an error of some 3−5° may be present at individual sites and could be systematic if a large number of sites occur in a region in which the tilt is uniform.

6

Statistical analyses

6.1 Introduction

An examination of lists of palaeomagnetic directions, poles, intensities, etc., can give an idea of their average value and scatter but statistical analysis allows the determination and testing of average directions and also affords measures of the scatter of individual observations. The method of calculating an average direction or pole position is outlined, followed by methods of analysing the accuracy of this average and the magnitude of the scatter of the individual observations. Statistical methods are described for analysing the intensity and susceptibility of groups of rocks and also the stability to partial demagnetisation of individual specimens. An assessment is then made of the two main assumptions behind the standard analysis of directions of magnetisation in rocks.

6.2 Average directions of magnetisations and pole positions

Directions of magnetisation are usually expressed in terms of their *declination,* which is measured clockwise from geographic north, and their *inclination* from the horizontal, denoted positive downwards and negative upwards (Figure 6.1). A direction of magnetisation can also be described in terms of the pole position towards which it is pointing in the same way as a compass points towards the present magnetic pole. The latitude and longitude of this pole position can be calculated (Figure 6.2) by assuming that the remanence was acquired in the field of a uniformly magnetised Earth. It is extremely important to realise that, at this stage, this pole position is merely another mathematical method of expressing the direction of magnetisation – any subsequent interpretation of these pole positions in

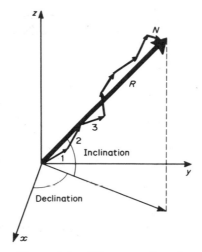

Figure 6.1   The mean direction

A number (N) of vectors (directions or pole positions) of unit length can be summed, using direction cosines, to obtain the direction and length of the resultant vector (R).

terms of past geomagnetic or rotational poles (Section 7.5) is a radically different conceptual step.

As remanent magnetisation is a vector (it has both magnitude and direction), standard vector algebra can be used to obtain the mean direction. Since there is no clear relationship between the intensity of magnetisation and the significance of the direction (Section 6.8), it is usual to give each vector unit weight, so that the declination (**D**) and inclination (**I**) of each direction can be expressed in cartesian coordinates by its direction cosines:—

$$x = \cos \mathbf{D} \cos \mathbf{I}; \quad y = \sin \mathbf{D} \cos \mathbf{I}; \quad z = \sin \mathbf{I}$$

These direction cosines can then be summed for a number of vectors to obtain the length of the resultant vector (**R**) and its direction ($\overline{\mathbf{D}}, \overline{\mathbf{I}}$):—

$$\mathbf{R}^2 = (\Sigma x)^2 + (\Sigma y)^2 + (\Sigma z)^2$$

$$\overline{x} = \frac{\Sigma x}{\mathbf{R}}; \quad \overline{y} = \frac{\Sigma y}{\mathbf{R}}; \quad \overline{z} = \frac{\Sigma z}{\mathbf{R}}$$

$$\overline{\mathbf{D}} = \tan^{-1} \frac{\overline{y}}{\overline{x}}; \quad \overline{\mathbf{I}} = \sin^{-1} \overline{z}$$

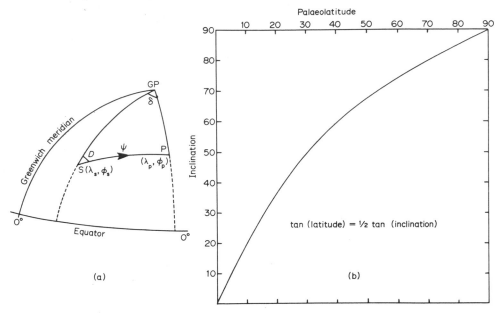

Figure 6.2   Calculation of pole positions

The site location (λ_s, ϕ_s) is known and the declination (D) relative to the meridian of the sampling site (ϕ_s) and the inclination (I). The distance of a pole from the sampling site, the colatitude, is given by $\psi = \cot^{-1}$ ($\frac{1}{2} \tan I$), so that the pole position (λ_p, ϕ_p) can be obtained graphically. Mathematically, the pole coordinates can be obtained from:

$$\lambda_p = \sin^{-1} (\sin \lambda_s \cos \psi + \cos \lambda_s \sin \psi \cos D$$
$$\phi_p = \phi_s + \delta \text{ where } \delta = \sin^{-1} (\cos \psi \sin D / \cos \lambda_p)$$

Pole positions can be put into cartesian coordinates and averaged in the same way, substituting longitude for declination and latitude for inclination.

The directions or poles can be illustrated graphically on a variety of projections. These are generally either orthomorphic (equal area) or zenithal (equal angle), depending on the need to illustrate the areal or angular relationship of the individual vectors. The centres of these projections are usually either polar or equatorial, the choice usually being made so that the vectors are plotted away from the circumference where the distortion is greater. Conventionally, the vectors are plotted as solid symbols on the lower hemisphere and hollow symbols on the upper hemisphere, i.e. in the polar projection of directions, positive inclinations plot as dots and negative inclinations as circles.

6.3   Estimates of precision and scatter of palaeomagnetic directions

Statistical analyses are required to estimate the reliability of the average direction and to measure the magnitude of the scatter of the individual directions. These two parameters can be defined in a variety of ways, but all depend on the relationship between the total number of directions (N) and the length of their average vector ($\mathbf{R}$).

(i)   Significance Points

A number of unit vectors drawn from a random collection would be highly unlikely to be identical ($N = \mathbf{R}$) or exactly antiparallel to each other ($\mathbf{R} = \mathbf{O}$) so that it is possible to calculate the probability (P) of particular vector sums ($\mathbf{R}$) occurring for various numbers of vectors (N) drawn from a collection. This probability can be determined for each mean vector but it is more convenient to compare the observed values with tabulated significance points, usually for either 95% or 99% probability (Figure 6.3). If the observed vector ($\mathbf{R}$) is the same length or greater than the corresponding significance point then the chances are 20 to 1 or 100 to 1 against the observed mean vector arising from that number (N) of random observations.

As with all statistical tests, the reliability is poor for low numbers, particularly below $N = 5$ and even using ten observations, instrumental variations of $\pm 1°$ cause changes in the second decimal place of the total vector length, resulting in appreciable changes in its apparent significance. The practical application of this test is therefore restricted to large numbers of observations.

(ii)   Fisherian Statistics

In response to the demand for a statistical model against which palaeo-magnetic observations could be tested, Sir Ronald Fisher developed a statistical model defining the distribution of points on a sphere. This model simulated the Gaussian (normal) distribution in three dimensions whereby points on a sphere (directions or poles) could be described in terms of a probability density (P) given by:

$$P = \frac{K}{4\pi \sinh K} \exp (K \cos \theta)$$

where θ is the angle between the observed individual directions and the true mean direction and K is the precision parameter, varying from 0 for a

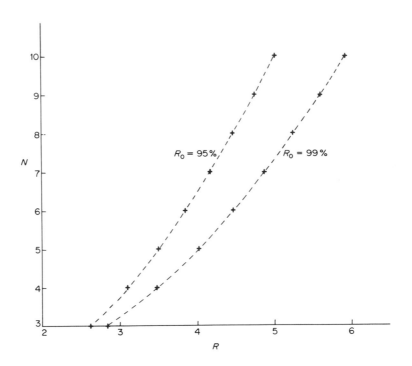

Figure 6.3   Significance points (R_0)
The significance points of 95% and 99% probability are shown graphically for up to 10 observations. The common approach of the lines at low N values is a measure of the sensitivity of the determination to small changes in the length of the resultant vector (R). Values R_0 for $10 > N \geqslant 20$ are given below and can be approximated for $20 \geqslant N \geqslant 100$ by:

$$R_{0_{95}} = (N \times 2.605)^{\frac{1}{2}} - 0.04$$
$$R_{0_{99}} = (N \times 3.781)^{\frac{1}{2}} \text{ (within } \pm 0.15 \text{ of actual value)}$$

N	$R_{0_{95}}$	$R_{0_{99}}$	N	$R_{0_{95}}$	$R_{0_{99}}$
11	5.28	6.25	16	6.40	7.60
12	5.52	6.55	17	6.60	7.84
13	5.75	6.84	18	6.79	8.08
14	5.98	7.11	19	6.98	8.33
15	6.19	7.36	20	7.17	8.55

perfectly random distribution, to infinity for identical directions. On this model, it is possible to estimate the *precision parameter*:

$$K \doteqdot k = \frac{N - 1}{N - R}$$

and the reliability of the observed mean direction can be defined by measuring the radius (α) of a circle on the sphere's surface, centred on the observed mean direction, within which there is a particular probability (P) of the true mean direction lying, the *cone of confidence*:

$$\alpha = \cos^{-1} 1 - \frac{N - R}{R} \left\{ P^{-1/(N - 1)} - 1 \right\}$$

In most palaeomagnetic studies, the probability, P, is taken to be 0.05 so that there is a 20 to 1 chance of the true mean direction lying within α_{95} degrees of the observed mean direction. (Circles for other probabilities can be measured, but this tends to be confusing). The two precision parameters, k and α_{95} can therefore be used as measures of the reliability of the observed mean direction of a group of directions or poles which have a Fisherian distribution, the highest reliability being for the largest k and the smallest α_{95} (Figure 6.4). As in all statistical analyses, these precision estimates are less reliable for low numbers and there is increasing uncertainty about the reliability of k particularly for N less than 7 and of α_{95} when k becomes less than 10. (The reliability of the statistical estimates can be checked by, for example, calculating their probable error ($P = 0.50$) e.g. in the estimate of k (Figure 6.5).)

The magnitude of the scatter of directions about their mean can be measured in an analogous way to that of the standard deviation in a Gaussian distribution. The definition of the Fisherian distribution involves θ, the angle between the observed individual direction and the mean direction so that for any Fisherian group of points on the sphere, defined by K, a circle of radius θ about the mean can be drawn which encloses a selected percentage (usually 63%, 95% or 50%) of the points:

$$\theta_{63} = 81 \, K^{-\frac{1}{2}}; \quad \theta_{95} = 140 \, K^{-\frac{1}{2}}; \quad \theta_{50} = 67.5 \, K^{-\frac{1}{2}}$$

In practice k, the estimate of K, must be used in this calculation. It is essential that a distinction is made between, for example, α_{95} and θ_{95};

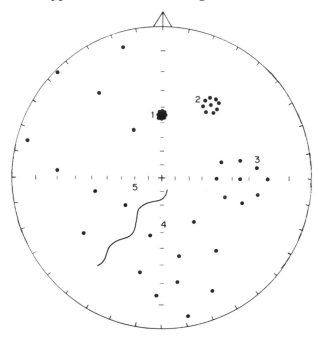

Figure 6.4   Models of scatter and statistical analysis

Five groups of nine points are shown above and the corresponding statistical analyses are shown below:—

Group	R	Significance	k	α_{95}	circular standard deviation
1	8.99117	0.00%	906	1.7	2.7
2	8.97695	0.00%	347	2.8	4.4
3	8.74997	0.00%	32	9.2	14.4
4	8.24349	0.00%	10.6	16.6	25.1
5	7.11617	0.01%	4.2	.28.4	40.0

the former is a measure of the accuracy of the mean direction and the latter is a measure of the scatter of directions about the mean. The circle enclosing 63% of the points is generally known as the circular standard deviation (θ_{63} = c.s.d.) by analogy with the Gaussian standard deviation (enclosing 68.3%) and can be used to derive the circular standard error (c.s.e.):

$$\text{c.s.e.} = \theta_{63}/N^{\frac{1}{2}}$$

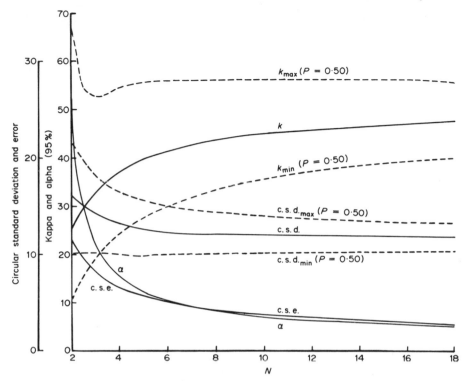

Figure 6.5   Statistical parameters for few observations

The parameters k, α_{95}, c.s.d., c.s.e. are shown for a scatter defined by $R = N \times 0.98$ for values of N between 2 and 18. The deviation of k and c.s.d. for $P = 0.50$ are shown as dashed lines indicating the low reliability of such estimates for values of N less than 6.

(A further, but non-Fisherian, measure of the scatter of directions is sometimes used, the angular standard deviation (δ) which is given by:

$$\delta = \cos^{-1} (R/N)$$

It is identical to the circular standard deviation if $N \approx N - 1$, i.e. for large numbers of observations, but differs at lower values. As this statistic is only strictly applicable to two-dimensional distributions, its significance at low N values is uncertain.)

A comparison can be made between the observed distribution of points on a sphere (directions or poles) and that of the Fisherian model (Figure 6.6).

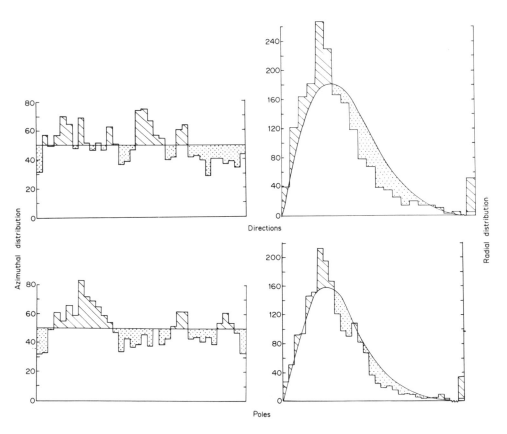

Figure 6.6   The distribution of directions and poles

The observed distributions are shown in comparison with the theoretical, Fisherian, distribution of azimuth on the left and angular distance from the mean on the right. Neither directions nor poles are strictly Fisherian, although the agreement is sufficiently close to warrant the use of Fisherian analyses in preference to other available statistical models.

This must be done in two parts and statistical tests applied to each part. The expected number of points (E) within an angular distance interval between θ_1 and θ_2 of the mean is given by:—

$$E = N \{ e^{-k(1 - \cos \theta_1)} - e^{-k(1 - \cos \theta_1)} \}$$

and the expected number of points distributed in azimuthal intervals around the mean, i.e. within $10°$, $20°$, etc., segments, should be constant $(E = N/\text{No.}$ of segments). The observed number of points (O) can then be compared with the expected number using the χ^2 test for intervals (I) in which $E \geqslant 5$:—

$$\chi^2 = \Sigma \frac{O^2}{E} - N \text{ for } (I - 3) \text{ degrees of freedom.}$$

If both χ^2 tests are positive, i.e. close to zero, then the observed points have a Fisherian distribution, but if either test is negative, then they do not have a Fisherian distribution at the probability level at which the test was carried out.

(iii)   Polar and directional 'errors'
Any Fisherian distribution of points can be adequately analysed using standard Fisherian statistics, but the geometrical relationship of directions and poles means that both cannot have a Fisherian distribution. It is therefore necessary to convert the parameters from the Fisherian to the non-Fisherian distribution. Converting α_{95} from a Fisherian distribution of directions, the error in inclination (δI) is independent of the distance from the mean pole, i.e. the average inclination $(\bar{I})$, but the error in declination (δI) is dependent on $\bar{I}$:

$$\delta D = \frac{\alpha_{95}}{\cos \bar{I}} \; ; \quad \delta I = \alpha_{95}$$

The circle of confidence (α_{95}) can therefore be converted to an ellipse of confidence around the mean pole nearest to the sampling site (Figure 6.7) using the directions $\bar{D} \pm \delta D$ and $\bar{I} \pm \delta I$ to define points around the pole. Similarly the length of the axes of the ellipse of confidence, centred on the average pole position, can be calculated along the mean declination (δm) and perpendicular to it (δp):

$$\delta m = \alpha_{95} \sin \psi / \cos \bar{I} ; \quad \delta p = \frac{1}{2} \alpha_{95} (1 + 3 \cos^2 \psi)$$

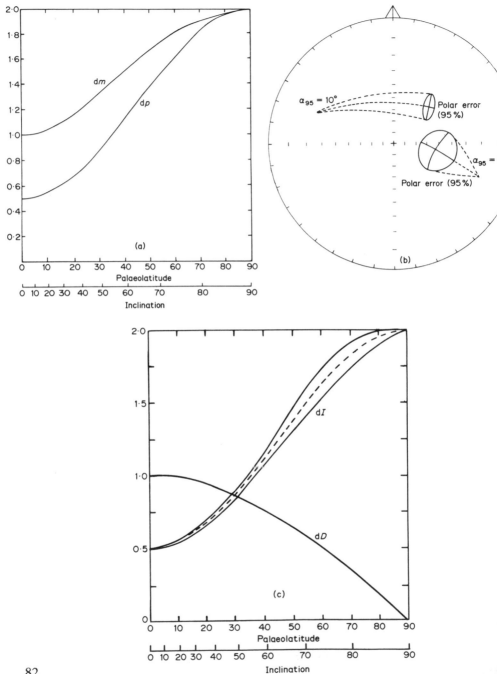

where ψ is the angular distance of the mean pole from the sampling location. (Strictly speaking, the ellipse is not quite centred on the pole, but the difference is small.) Similar considerations apply to the conversion of other statistical parameters derived from a distribution of directions to the different distribution of poles.

6.4   Analysis of groups of directions

When the factors influencing the distribution of directions are complex, as in the case of the remanence in rocks, it is useful to be able to isolate the various components which contribute to the total scatter. Conversely, it is often desirable to combine groups of already analysed directions, but this can only be done simply if the groups have been taken from the same overall population of directions.

If groups of directions have been taken from the same overall population then their directions and precisions must be significantly similar. If there are B groups, each containing N_i observations and with an average·vector length and precision of $\mathbf{R}_i$ and k_i respectively, and the total length of the sum of all observations is $\mathbf{R}_B$, then the agreement between precisions and directions can be tested using the standard statistical table for F-values. The precision of pairs of groups can be tested using $F = k_1/k_2$ for $2(N_1 - 1)$ and $2(N_2 - 1)$ degrees of freedom. The similarity of all groups of directions can be tested simultaneously using:—

$$F = \frac{\Sigma N_i - B}{B - 1} \cdot \frac{\Sigma \mathbf{R}_i - \mathbf{R}_B}{\Sigma N_i - \Sigma \mathbf{R}_i}$$

For the groups to be statistically identical, both F-tests must be positive and the average precision (k_B) for the collection as a whole is then given by:

$$\frac{1}{k_B} = \frac{1}{k_1} + \frac{1}{k_2} + \frac{1}{k_3} \ldots \ldots$$

Figure 6.7   Polar 'errors'
(a, b) The statistics of a Fisherian group of directions can be converted to correspond to the distribution of the corresponding poles. These statistics, such as α_{95}, result in an ellipse around the mean pole determined by the axes δm and δp, which vary according to the palaeolatitude of the sampled area. (c) An actual error in declination, δD, or inclination, δI, arising for example from an error in orientation also has an effect on the pole position which is dependent on the palaeolatitude of the sampled area.

It is possible to distinguish scatter due to instrumental and orientation errors by analysis at different levels — specimen, samples and sites (Section 6.5) but there is generally a significant within-site scatter despite the averaging effects of such analyses. This is because the between-site scatter is formed by three components — changes in the field during the formation of the different sites, bedding (tilt) orientation errors and the remaining within-site scatter. These are, of course, largely averaged out in obtaining the overall average direction, but the magnitude of the between-site scatter is important in studies of the time variations of the past geomagnetic field (Section 7.3), and if the estimate of precision at each site (ω) is approximately constant, allowance can be made for this effect on the estimate of between-site precision (β):

$$\omega = \frac{\Sigma(N_i - 1)}{\Sigma(N_i - R_i)} \; ; \quad \beta = \overline{N} \left\{ \frac{\Sigma R_i - R_B}{B - 1} - \frac{1}{\omega} \right\}^{-1}$$

where $\overline{N}$ is weighted average number of directions per site ($\overline{N} = (\Sigma N_i - \Sigma N_i^2)/(B - 1)$). In practice, the within-site dispersion is variable and the assumption of similarity should be checked by an F-test on the site dispersions. If the within-site scatter is equal or greater than the between-site scatter, then clearly the effect of field changes or tilt errors are less than the factors affecting the scatter within a site, and it is therefore necessary to test that the two precision estimates are significantly different from each other:

$$F = 1 + \frac{2\overline{N}\omega}{\beta} \quad \text{for } 2(B - 1) \text{ and } 2\Sigma(N_i - 1) \text{ degrees of freedom.}$$

6.5   Levels of statistical analyses

Statistical analyses can be carried out at different sampling levels (specimens, samples, sites, etc.), different demagnetisation levels, and various combinations of these. The precision of the statistical analysis is dependent on the number of observations and this suggests that the choice should involve the maximum number, but the statistical parameters are determined by specific physical properties, such as the changing direction of the previous Earth's field. As the reason for the analysis is usually to investigate the physical properties, the method of analysis must be determined by these considerations rather than by purely numerical factors.

Samples from a single 'site' are considered to have acquired their primary magnetisation simultaneously. Averaging specimen values to obtain sample values tends to reduce the effect of instrumental and inhomogeneity effects. Similarly, averaging from samples to sites tends to average out orientation errors. These processes therefore allow some estimate to be made of the magnitude of different types of error so that observations which involve substantial errors can, at each stage, be rejected from the analysis on objectively defined physical characteristics. If the number of specimens per sample is constant then there is little difference in the average direction or the measures of scatter whether calculated on sample or specimen observations unless individual specimen directions are markedly anomalous. If the number of specimens per sample is uneven, then, because of the use of unit weight per vector, the average values are weighted towards the more densely studied samples and there is less tendency to average out orientation errors. If all samples at the site were not in fact magnetised simultaneously, as in the case of some sedimentary rocks, all these considerations must be re-examined. However, such an average site direction, no matter how it is calculated, will average out the effects of time variations in the field during the acquisition of its magnetisation, so that, although the average field direction calculated from several different sites will still be valid, the scatter of the individual site directions will be less than the actual variations of the field.

Incremental demagnetisation is undertaken to isolate the primary magnetisation from remanence acquired later and is ideally carried out on all specimens as the magnetic properties of rocks can vary substantially within the same site. The most stable remanence, i.e. that remaining when the secondary components are minimised and spurious components (such as anhysteritic or chemical magnetisation) are insignificant, can be determined subjectively for each specimen and these can then be averaged. In large collections, this detailed analysis is impractical and one of two methods can be adopted. Pilot specimens can be selected which are thought to be typical of either each site or each sample and their stability can be investigated in detail (Section 4.5). This detailed information can then be used to decide the optimum treatment for each sample, each site or the entire collection. The remaining specimens can then be treated at their appropriate level and the directions of most stable remanence isolated in this way can then be combined statistically. Alternatively all specimens can be demagnetised at a small number of different levels and the optimum level at which the individual stable components are best isolated must then be objectively established. As detailed information is usually lacking at specimen and sample levels, it is generally assumed that the properties within the same site are comparable so that the level at which the remanence is purest (i.e. when there are the least secondary or spurious components) is achieved when the specimen or sample directions in a site are most closely grouped together. The optimum level of treatment for each site is therefore defined as that producing the maximum site precision (k). Site observations at their optimum level can then be combined and analysed.

6.6   Measurements of stability

The most common stability test is the demagnetisation process, designed to remove low coercivity components in order to isolate a single stable component which can be attributed to a field direction at some specific time (Sections 4.5 and 4.6). This process usually involves the introduction of some spurious magnetic components, particularly at high magnetic fields (Section 5.5) but there is normally a range of treatment when the low coercivity components are minimal and spurious components are still insignificant. The limits of this range and the recognition of the isolated component within it are usually subjective evaluations, terms such as 'highly stable', 'poorly stable', 'unstable', being used which are variable and therefore unsatisfactory. Two objective measurements have been devised to allow direct comparisons of the stability of remanence of different specimens to incremental demagnetisation by either alternating magnetic fields or temperature rises.

The *Stability Factor* (S.F.) is based on changes of direction and intensity during incremental (step) demagnetisation and is given for each individual step (i):

$$\text{S.F.}_{\cdot i} = \mathbf{R}_i/(\mathbf{R}_i + \Sigma r)$$

where $\mathbf{R}_i$ is the vector at step i and Σr is the non-vector sum of all previous changes from one step to the next. In practice, the stability factor is therefore mainly a measure of the ratio of low and high coercivity vectors and, as such, is a good measure of the physical stability of remanence.

The *Stability Index* (S.I.) is based on an evaluation of the closest possible grouping of directions which can be distinguished, irrespective of their intensity, over the widest possible range of thermal or alternating field treatment. The index is calculated for all possible combinations of three or more successive directions during the demagnetisation process, and the most stable range is that for which the index is maximum:

$$\text{S.I.} = \max \{\text{Range}^{1/2}/\text{c.s.d.}\} \text{ of three or more successive directions.}$$

The most stable range is therefore defined and the directions which have been measured within that range can then be evaluated by standard Fisherian analysis. This index is suitable for determining the range over which a single component of remanence is best isolated and also allows an objective measure of the degree of its isolation from other components. It is therefore more directly applicable to palaeomagnetic studies than the stability factor, which is more suitable for rock magnetic studies.

Both measures must be used carefully and can only be applied when a high coercivity component can be isolated from other components. If, for example, the remanent vector changes from its low coercivity direction towards a stable higher coercivity direction, it may not, in fact, reach the stable direction before it becomes swamped by the addition of spurious components. Under such circumstances, both measures would indicate poor stability although it may still be possible to isolate the stable component by other techniques (Section 4.5).

6.7   Susceptibility and intensity of magnetisation

The susceptibility of rocks varies according to the amount, form and composition of the magnetic minerals which they contain and is generally in the range of 10^{-5} to 10^{-7} G Oe^{-1} in haematite bearing sediments and between 10^{-3} to 10^{-4} G Oe^{-1} in basic igneous rocks. Although the absolute values of susceptibility vary from one rock type to the next, the distribution of susceptibility in any one rock type is usually similar (Figure 6.8), i.e. with a peak at low values tailing off towards high values. This is a log-normal form of distribution, i.e. the logarithms of the values have a symmetrical (Gaussian) distribution. In such a distribution, the average value ($\bar{M}$) and the

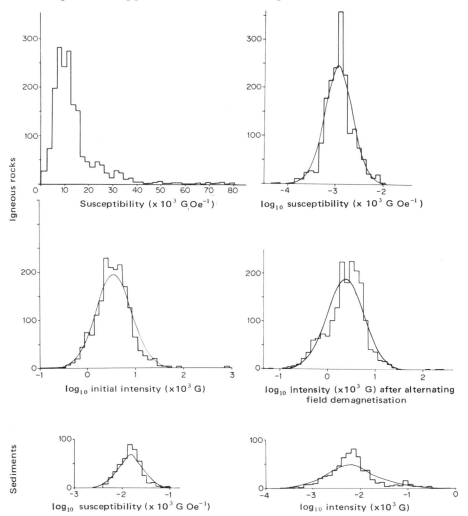

Figure 6.8   The distribution of intensity and susceptibility in rocks

The top left graph shows susceptibility values, for a typical igneous formation, plotted on a normal scale, and the top right shows the same values on a logarithmic scale. The log normal curve is shown and the agreement between the observed and theoretical distribution is very close. Similarly the intensity of remanence in igneous rocks before and after partial demagnetisation (central graphs) show close agreement and the susceptibility and intensity of remanence of sediments is also log normally distributed (lower graphs). (Sedimentary data from Irving *et al*, 1966.)

standard deviation (s) or other measures of the scatter should be calculated on the logarithms of the measured values:

$$\overline{M} = \log^{-1} \frac{\Sigma_1^N \log x}{N} \qquad s = \log^{-1} \pm \left\{ \frac{\Sigma_1^N (\log \overline{M} - \log x)^2}{N - 1} \right\}^{1/2}$$

where there are N specimens, each of susceptibility x.

The intensity of remanence is controlled by many factors. In sedimentary rocks, the intensity of natural remanence is generally up to 10^{-5} G, and in igneous rocks up to 10^{-2} G, although much higher intensities can occur, particularly if the rocks have recently been struck by lightning. The intensity distributions in different rock types show similar log-normal type distributions (Figure 6.8) so that the statistical parameters should be calculated on the logarithmic values. This log-normal distribution suggests that the main influence on both susceptibility and intensity of remanence within any rock type is the distribution of grain sizes as these are log-normally distributed. In detail, susceptibility values tend to be closer to the pure log-normal distribution than the intensity values although both are usually leptokurtic (more tightly grouped around the average than a pure distribution), but during demagnetisation the intensity distribution tends to become much closer to the pure log-normal curve.

Direct comparisons can be made between susceptibility values in one formation and that of another, but the intensity of the Earth's magnetic field varies from the Equator to the Poles (Section 7.2) and so the intensity of primary and viscous remanence of rocks varies according to the latitude in which they were acquired. Intensity values must therefore be corrected to a common datum before direct comparisons can be made. As there is general evidence that, on average, the geomagnetic field has been dipolar for most of the Earth's history (Section 7.5), the strength of the field (F) in which the remanence was acquired can, for each collection, either be converted to its value at some particular latitude, such as the Equator (F_0), or it can be used to compute the past magnetic moment (M) of the Earth:

$$F_0 = F (1 + \sin^2 \lambda)^{1/2}$$
$$M = F/(1 + 3 \sin^2 \lambda)^{1/2}$$

where λ is the palaeolatitude (Figure 6.2). Differences between the corrected intensity values must reflect either real differences in the intensity of the field or, of course, that the Earth's field was not at that time truly dipolar.

6.8   Discussion

Two main assumptions are involved in the present statistical analysis of palaeomagnetic data. The first is that every vector has equal significance and is therefore given equal weight in the analysis. The second is that because Fisherian statistics were developed in part to meet the requirements of palaeomagnetism, it is assumed that directions or poles have a Fisherian distribution.

Clearly different remanence vectors have different significance as some can be demonstrated to be more stable or more precisely defined that others. Unfortunately, many palaeomagnetic results consist of relatively few observations and analyses are therefore carried out at a low level of statistical reliability. In such cases it is difficult to use a satisfactory weighing system based on such criteria as stability (although the stability index may be developed for this purpose), precision of the average sample or site vectors, etc. A very coarse method of weighting is often applied in which certain directions or poles are entirely omitted from the analysis. This is obviously acceptable when it can be shown that the rejected observations are meaningless because of instability, inhomogeneity, etc. It would obviously be preferable if a proper weighting system could be developed using definable physical characteristics. The intensity of remanence has sometimes been used as a weighting factor and this is an acceptable method if there is a clear relationship between the intensity and the precision of measurement of the vector, e.g. if the measurements were carried out at the lower limits of instrumentation. Generally, however, there is no obvious reason for considering that a low intensity vector is less meaningful than a higher intensity vector.

Few tests have been done to investigate whether directions or poles have true Fisherian distributions, mainly because rigorous testing is difficult owing to the large numbers of observations required. However, such tests show that, in detail, neither directional nor polar distributions are truly Fisherian either in azimuth or distance from the mean. However, the differences, although statistically significant, are small and do not suggest that any serious error is introduced by assuming a Fisherian distribution for most palaeomagnetic data. The observed vectors, in fact, tend to be slightly closer to their mean than they would on the theoretical distribution so that the statistical parameters slightly overestimate the scatter and slightly underestimate the precision. A weighting system, which emphasises the observations near the mean, would cause the observed distribution to become excessively peaked (leptokurtic) compared with the Fisherian model, and such a distribution would be more analogous to a three-dimensional log-normal distribution, for which new statistical parameters would need to be derived.

7

Geomagnetic applications

7.1 Introduction

Measurements of the Earth's magnetic field have been made regularly in specific locations from about 1600, although most continuous records are only available for this century. These direct observations have allowed a detailed picture of the present field to be determined (Section 7.2) but are insufficient for studying long-term variations ($\sim 10^3$ years). Study of the remanent magnetisation of rocks affords the facility for extending measurements of the geomagnetic field throughout geological time. In particular it is possible to determine the variation in direction (Section 7.3) and intensity of the field (Section 7.4) and the overall pattern of the field (Section 7.5). These studies, however, depend on a knowledge of the age of a remanence which must be an accurate reflection of a previous geomagnetic field. It is necessary, therefore, for the material under study to be isotropic and homogeneous and for a sufficient number of measurements to be made to ensure averaging out of instrumental errors (Section 5.4). It is particularly important that allowance is made in the analysis for possible errors in the tilt correction (Section 5.7). Under these circumstances, the geomagnetic field direction should be definable within some 2° at each site. However, systematic reductions in magnetic inclination have occasionally been suspected within sedimentary sequences compared with the inclination in contemporaneous igneous rocks. This discrepancy, if genuine, may represent compactional effects occurring during diagenesis, rather than an actual depositional 'error' (Section 4.2). It is, however, possible to test for this reduction of inclination by comparison of the pole positions using both declination and inclination values (Section 6.2) with those determined from the intersection of meridian lines (Section 7.5). In general, however, such discrepancies are so small that they lie within the statistical limits with which the mean direction of remanence is defined.

The geomagnetic applications of palaeomagnetism are clearly of importance in determining the evolution of the Earth's magnetic field and therefore of the Earth itself, and studies of extraterrestrial material are expanding this information to a better understanding of the evolution of the solar system.

7.2 The present geomagnetic field

Magnetic observatory records of the directional components of the intensity of the Earth's field show 12 and 24 hour, monthly and annual cycles which differ from one observatory to the next, but rarely exceed 0.1% of the average field strength. These detailed records also show interruptions by magnetic storms, sometimes as strong as 1% of the field, and persisting for several days. These are caused by bursts of solar radiation and show an 11 year cycle. All these variations are directly related to rapidly varying electrical currents which circulate in the ionosphere of the Earth as a result of radiation from the Sun. These ionospheric currents have a direct magnetic effect at the surface of the Earth and also induce secondary electrical currents with associated magnetic fields within the Earth itself. The origin of these short term variations however, is essentially external to the Earth and it is necessary to average out corresponding transient variations from the observatory records, in order to obtain the intensity and direction of the geomagnetic field freed from the external components. The observatory data alone are insufficient to define the total pattern of the Earth's field and are supplemented by regional surveys which are corrected for transient variations recorded at the nearest magnetic observatory. This data is then plotted and smoothed to reduce the effects of local magnetic anomalies, and charts of the Earth's field are compiled, usually at 5 yearly intervals.

The two main features reflected by these charts (Figure 7.1) are the dipolar nature of the Earth's field (i.e. the two positions where the inclination (dip) of the field is vertical), and the systematic change of direction and intensity between the two poles (the intensity is about 0.6 Oe at each dip pole and about 0.3 Oe at the equator). The present *magnetic (dip) poles* are at approximately 73°N 100°W and 68°S 143°E. They are not antipodal but correspond closely to an inclined, eccentric

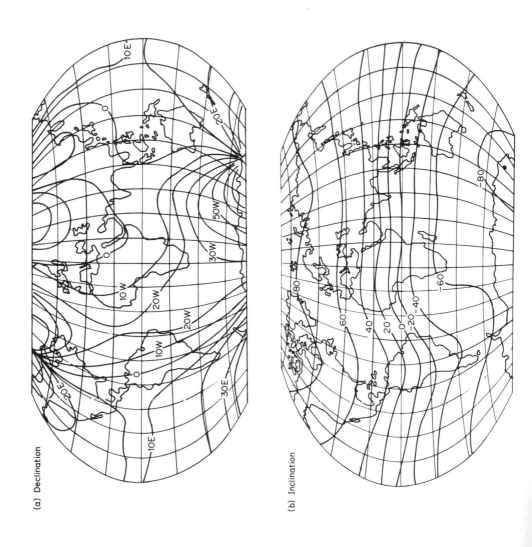

(a) Declination

(b) Inclination

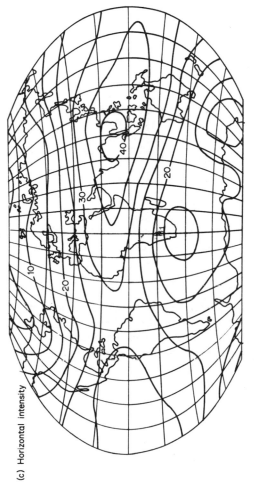

Figure 7.1   The present Earth's magnetic field (1955)
(a) The declination (variation) of the horizontal vector from the geographical meridian.
(b) The inclination from horizontal (magnetic dip). (c) The horizontal intensity of the field (0.05 Oe intervals).

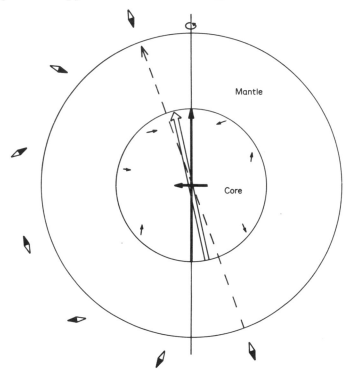

Figure 7.2   Models of the Earth's magnetic field
The dashed line marks the orientation of the inclined, non-geocentric dipole passing through the two dip poles on the Earth's surface and the compass needles around the edge are shown inclined in relation to this field. The field can be modelled mathematically using axial and equatorial dipoles to define the main inclined dipole (hollow arrow) which forms some 80% of the total field observed at the Earth's surface. The non-dipole portion of the field can be interpreted in terms of a number of small dipoles near the core-mantle interface.

magnetic dipole situated some 400 km away from the centre of the Earth (Figure 7.2). The field can be more precisely defined in terms of spherical harmonics, i.e. by determining the amount of the field which can be attributed to the potential of variously oriented hypothetical dipoles at the centre of the Earth, and the amount attributable to external sources.

Spherical harmonic analysis shows that more than 99.5% of the Earth's field is of internal origin; the balance probably reflecting uncertainties in the

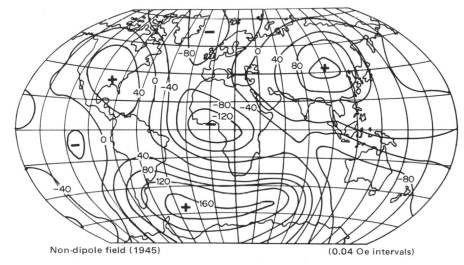

Non-dipole field (1945)   (0.04 Oe intervals)

Figure 7.3   The non-dipole field (1945).

estimated field in parts of the Antarctic rather than any significant external contribution. The analysis also shows that some 80% of the Earth's field can be attributed to a single geocentric dipole (Figure 7.2) inclined at $11\frac{1}{2}°$ to the Earth's axis of rotation with a magnetic moment of 8×10^{25} G and its axis intersecting the Earth's surface at the *geomagnetic poles* (78.5°N 69.1°W, 78.5°S 110.9°E). Most of this dipole field ($\sim 80\%$) is attributable to a single dipole along the Earth's axis of rotation, and the remainder to a dipole or dipoles in the Earth's equatorial plane. When the inclined dipole field is subtracted from the total field, the remaining *non-dipole field* (Figure 7.3) shows some eight regions, of continental dimensions, displaying either positive or negative values with an amplitude of some 0.15 Oe. These regions are assymmetrically distributed, are particularly strong in the southern hemisphere, and notably weak in the Pacific area.

Few direct, continuous records of the Earth's field go back more than a 100 years or so, since most magnetic observatories were only established during this century. However, comparisons of the available observatory records (Figure 7.4) or magnetic charts show that the field has gradually changed in both intensity and direction. These long term changes, *secular variations,* differ at each location and are caused by a growth and decay of

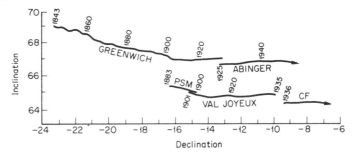

Figure 7.4   Observatory records of secular variation in London and Paris
Earlier records, not shown on this diagram, are less reliable but suggest an elliptical motion of the geomagnetic vector. PSM = Parc St. Maur   CF = Chambon la Forêt

the field pattern, superimposed on a westerly drift of the non-dipole field of some 0.2–0.3 degrees of longitude per year. This rate of westerly drift appears to change with alteration in the Earth's rate of rotation. In contrast to the drifting non-dipole field, the main dipole appears to have been approximately constant in orientation since Gauss made the first spherical harmonic analysis in 1839 (a postulated 4° westerly drift between 1839 and 1880 is probably due to insufficient data in earlier analyses) and its intensity appears to have been decreasing fairly constantly at a rate of about 5% per century. If this rate of decrease continues unchanged the main dipole field will disappear in a further 2,000 years time. Routine satellite observations now supplement the information on the Earth's present field and regular surveys are now undertaken in areas of the world which were previously unmapped. Therefore the main advances in our understanding of the origin of the Earth's magnetic field depend not so much on more precise mapping of its present pattern, but on extending the record through past geological time. In particular, it is necessary to determine if the present asymmetries of the non-dipole field are permanent and if the overall dipole nature and intensity characteristics of the Earth's field have persisted over geological time.

7.3   Secular variation

Records of long term variations ($> 10^2$ years) of the Earth's field can be extended back for a few thousand years using the remanence preserved in historically dated lavas and archaeological material. Archaeological samples,

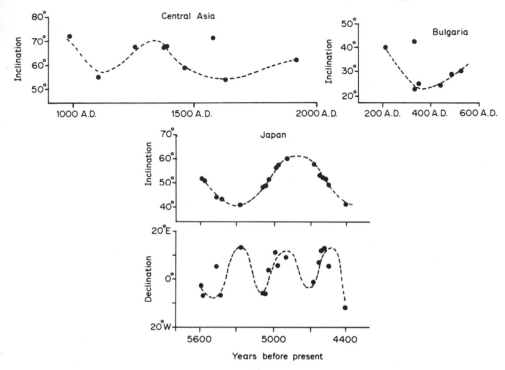

Figure 7.5 Secular variation from archaeological and historical observations
using lavas, pottery, bricks and kilns
(After Watanabe, 1959 and Burlatskaya *et al*, 1965).

such as kilns, furnaces, tiles, pottery, etc., are often ideal for palaeomagnetic
work, as the processes by which they acquired their thermal remanence can
be duplicated in the laboratory, and their magnetic constituent is usually
haematite, which is magnetically and chemically stable. Using these
materials, it is possible to determine the declination, inclination and
intensity (Section 7.4) of the field at particular places for specific times, and
so construct a graph of field variations during the last few thousand years
(Figure 7.5). The main difficulty is the precise dating of samples for which
there are no historical records. Samples from kilns and furnaces can often be
isotopically dated within ±100 years by the C_{14} method, using ashes
remaining from their last firing, but an inherent limitation of this technique
is that radiogenic C_{14} is produced by cosmic radiation, the rate of incidence

of which fluctuates at the Earth's surface with changes in the strength of the geomagnetic field. Tiles and pottery can often be dated archaeologically, but these materials only afford observations of the inclination and intensity of the historic field as there is no record of their orientation within the furnace. Precisely dated historical and archaeological data are now available from several continents and usually show a periodicity of secular variation of about 10^3 years and an amplitude of some $20°$ (Figure 7.5); a picture broadly consistent with modern observatory records.

For earlier times, all methods of absolute dating become too imprecise for the detection of age differences of several thousand or even millions of years. Relative dating, however, can be used in special circumstances, for example, on glacial varve deposits and certain lava sequences. Varves are sediments which show alternating bands of coarse and fine deposits, the coarse sediment having been deposited during summer thaws and the finer particles having settled out during winter freezing, so that the relative age of varves within a sequence can readily be determined. The remanence of glacial varves (Figure 7.6) deposited during the last 100,000 years or so seem to reflect a

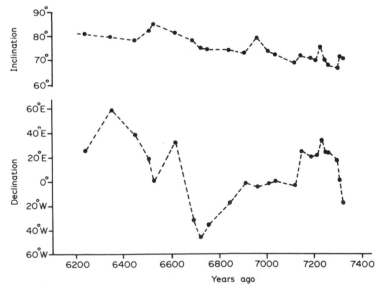

Figure 7.6   Secular variation using varve deposits
(After Granar, 1958).

similar periodicity and amplitude of secular variation to that observed in archaeological material, although the magnetic stability of most available data is still somewhat uncertain. (Varve deposits are also known for very much earlier periods, but most of these have been indurated and their original depositional remanence has been modified by subsequent chemical changes (Section 4.4).) Lava eruptions are obviously irregular but are often restricted to certain parts of a volcano for prolonged periods so that some lava sequences have accumulated during a single secular variation cycle (Figure 7.7). However, since the interval between successive lavas cannot be precisely determined it is not possible to measure the periodicity or pattern of past secular changes by this means. Nonetheless, each lava preserves the direction of the geomagnetic field at a single precise time, and the scatter between site directions in a uniform lava sequence is attributable to a com-

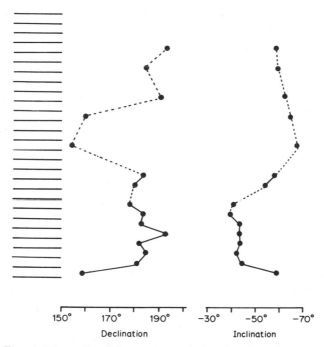

Figure 7.7   Possible secular variation in a lava sequence
The lavas are thought to have been erupted during the course of some 500 years about 15 million years ago in Oregon, U.S.A. The solid line links directions measured in adjacent lavas. (After Watkins, 1965).

ponent of within-site scatter (mainly due to instrumental and orientation errors) and the secular changes in the field between the eruption of different lavas (Section 6.3). It is sometimes possible to isolate the between-site scatter of directions from the effects of the within-site scatter (Section 6.4) and obtain a direct measure of the average magnitude of secular change, irrespective of the number of cycles involved. Between-site scatter observed in sedimentary rocks and attributed to secular changes must represent minimum estimates and therefore be treated with caution because of the possibility of long-term chemical magnetisation subsequent to their deposition.

The significance of the observed scatter of site directions, or their corresponding poles, can be assessed by comparing them with various models of the secular variation of the Earth's field. There are three basic models for the behaviour of the main (inclined) dipole field. In the simplest of these the dipole remains inclined but drifts westwards with the non-dipole field. In the other two models the inclination of the main dipole varies as a result of a changing ratio of the axial and equatorial components. Depending on the way in which the equatorial dipole changes, the main dipole either oscillates along various lines of longitude continually passing through an axial geocentric position or undergoes a form of 'random walk' which is centred on the axis of rotation. The pole positions on all three models have a symmetric circular distribution centred on the axis of rotation, and this distribution is independent of the latitude of the sampling sites. Conversely the scatter of corresponding mean site directions of magnetisation is latitude dependent and the shape of their distribution varies from circular near the poles to oval at the equator. Superimposed on models of the main dipole field must be various models of the changing non-dipole field; most of these imply some latitudinal dependence for the distribution of pole positions and sometimes of site directions.

The crucial test between these models depends on determining the shape of the distribution of either site directions or their corresponding pole positions. Unfortunately insufficient data are available to test the shapes of the distributions except for a few detailed collections and these are mostly from rocks less than 70 million years old. However, such tests suggest that the model of a westward drifting main dipole field is unlikely as the pole positions in this case would have an annular distribution around the axis of rotation, whereas the observed distributions are in fact strongly centred (Section 6.8 and Figure 6.6). Comparisons of the magnitude of scatter of

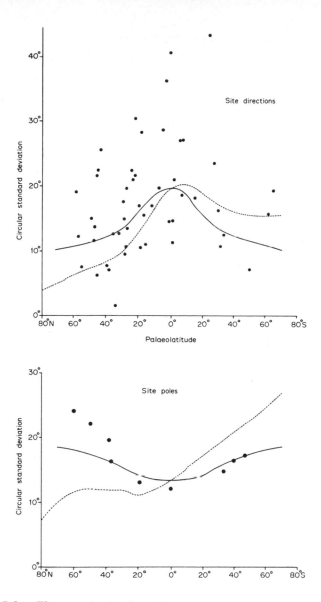

Figure 7.8   The magnitude of secular variation from igneous rock studies
The circular standard deviation of directions or poles is a measure of the magnitude of
secular variation and is shown as a function of the palaeolatitude in which the rocks
formed. This can be compared with the scatter of the Earth's magnetic field directions or
poles along present day latitudes (dotted line) and also after averaging (solid line) for differences between the northern and southern hemispheres. The data are for 10 or more sites
but are of variable reliability.

either site directions or poles of different ages suggest that both show a dependence on latitude (Figure 7.8). This latitudinal dependence appears to be independent of the hemisphere, i.e. the variation in magnitude is symmetric about the equator, suggesting that the present assymmetry of the geomagnetic field between the hemispheres is a temporary phenomenon. However, abnormally low magnitudes of scatter observed in palaeomagnetic studies suggest that the present reduced rate of secular change in the Pacific region has been a permanent feature during the last million years and possibly during the last 5 million years.

7.4   The intensity of the ancient field

The absolute strength of the ancient field (F_{ANC}) can be determined by a comparison of the intensity of natural remanent magnetisation (M_{NRM}) with the intensity of thermal remanence (M_{TRM}) acquired by an igneous rock or baked archaeological sample during heating and cooling in a known field (F_{LAB}):

$$F_{ANC} = M_{NRM} \quad F_{LAB} / M_{TRM}$$

provided that there has been no physical or chemical change since the rock acquired its original magnetisation. Unfortunately most samples have acquired a viscous remanence and have often undergone physico-chemical changes during laboratory heating. The effects of viscous components can be overcome by comparing only the stable, high temperature components of natural (M_{PNRM}) and thermal remanence (M_{PTRM}) using partial thermal demagnetisation:

$$F_{ANC} = M_{PNRM} \quad F_{LAB} / M_{PTRM}$$

Similarly, comparison can be made between stable components of natural and thermal remanence isolated by alternating magnetic field demagnetisation. This technique is more subject to instrumental error (Section 5.5), particularly at high fields, but has the advantage of avoiding the repeated heating and cooling of a specimen, and the consequent risk of chemical changes occurring.

When there has been no physical or chemical change in the specimen during the experiment, the ratio PTRM : PNRM remains constant and the ancient field strength can then be determined (Figure 7.9). Unfortunately

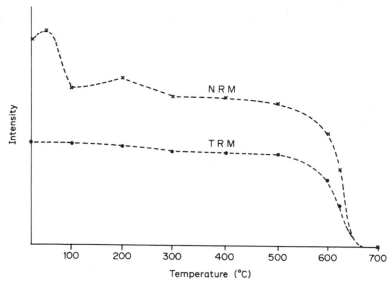

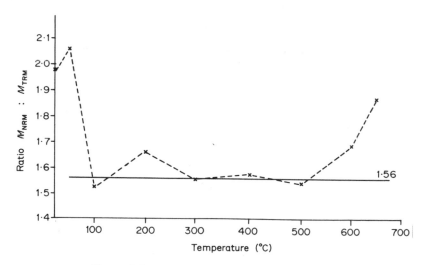

Figure 7.9   Ancient intensity determinations

Comparison of the intensity of natural and thermal remanence acquired in a known field shows that this specimen must have acquired its original remanence in a field which was 1.56 times greater than the field applied in the laboratory when giving it a thermal remanence.

some changes usually occur when the specimen is given a thermal remanence and precise field determinations are then impossible although the approximate magnitude of the field can be evaluated if the ratio remains constant within about ±20%. The largest uncertainty is, therefore, the degree of natural physico-chemical alteration which the rock has undergone since it originally acquired its remanence.

Studies of archaeological material and historical lavas (Figure 7.10) show that the magnitude of secular variation of field strength is approximately 10−15% of the average field. Studies on older rocks (Figure 7.11) are inadequate in quantity and quality to determine the magnitude of secular

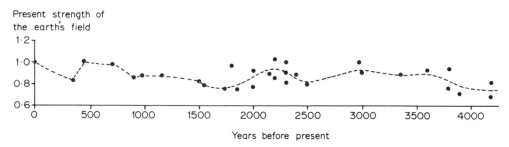

Figure 7.10   Secular variation of intensity in India using archaeological materials (After Athavale, 1966).

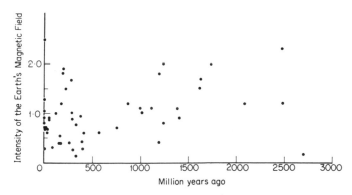

Figure 7.11   Intensity variation of the intensity of the Earth's field during geological time

The data have been corrected for palaeolatitude but not secular variations. All available determinations (1970) are shown, irrespective of the reliability of the determination or the precision of the age.

change at earlier periods so that, although the available observations have been corrected for their palaeolatitude (Section 6.7), they are uncorrected for secular variations. Observations on very old rocks ($\sim 10^9$ years) are based on natural remanence usually carried by haematite which, in some cases, has developed from the disintegration of other minerals (Section 3.2) so that the age of remanence and therefore the field is uncertain. Similar uncertainties exist about some younger rocks, but the general consistency of the observations suggests that the strength of the Earth's field has not changed substantially throughout geological time, with the possible exception of a reduced field strength during the Palaeozoic (300–600 million years ago).

Sedimentary rocks cannot normally be used for absolute determinations of the strength of past magnetic fields as their remanence is usually at least partly chemical in origin. However, where the remanence is entirely of depositional origin, its intensity is proportional to the number of aligned particles, which is itself dependent on the strength of the ambient field. When the sediment is redeposited in a known field, the intensity of laboratory remanence and natural remanence can be compared and the past field strength deduced. In general, however, only recent sediments can be used for such experiments as older rocks have usually been cemented and chemically altered since deposition.

Relative intensities of the ancient field can be rapidly determined by comparing ratios of natural remanence and susceptibility (the Q ratio). If samples have not been altered physically or chemically, variation in the Q ratios reflects changes in the strength of the field in which the remanence was acquired. Such comparisons are clearly more meaningful after the removal of secondary components or when using samples which have had a similar geological history. This technique is most useful for studying variations in field strength over a few thousand or few million years, such as a period of polarity transition (Section 8.5). During such transitions the Q ratios appear to be consistently lower than in adjacent polarity zones, which suggests that the Earth's field strength is reduced in these intervals.

7.5 The average nature of the ancient geomagnetic field

The average pattern of the geomagnetic field for historical and archaeological times can be obtained from analyses of site directions from different parts of the world. When the pole positions corresponding to these site directions are

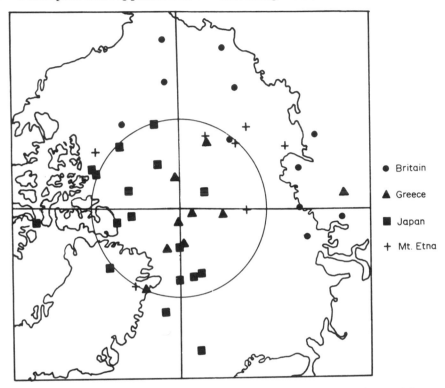

Figure 7.12   Pole positions determined from archaeological material
The observations represent a few thousand years of historical and prehistorical times and are somewhat unevenly distributed within each area, but, taken as a whole, are clearly centred on the Earth's present axis of rotation.

plotted (Figure 7.12), their distribution is clearly centred on the Earth's axis of rotation, suggesting that during the last few thousand years, the average field has been that of an axial geocentric dipole. This picture can be extended back for a million years or so using the remanence of deep-sea sedimentary cores. These cores are usually 15–30 m long and penetrate sediments deposited during the last 2–3 million years. Specimens from cores usually represent some 5–10,000 years of deposition and secular variations are therefore largely averaged out within single specimens. The cores available from most oceans are usually unoriented in azimuth, but the inclination of remanence is the same as that expected for an axial geocentric dipole field (Figure 7.13).

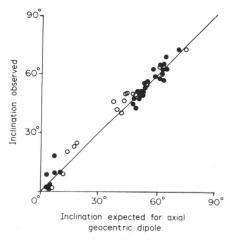

Figure 7.13   The inclination in deep-sea sedimentary cores
The sediments, generally less than 2×10^6 years old, show inclinations which are statistically identical to those expected for an axial geocentric dipole field during this period. (After Opdyke and Henry, 1969.)

In data from older rocks, tectonic movements become increasingly significant, but igneous lavas, less than 20 million years old, can often be assumed to have undergone little movement since their eruption. The pole positions corresponding to over 1,000 individual lavas of this age range (Figure 7.14) are clearly centred around the Earth's axis of rotation. There is, however, a distinctly uneven distribution of sampling sites, since the bulk of these observations are from comparatively few locations in the northern hemisphere. These observations clearly indicate that the geomagnetic field has been, on average, an axial geocentric dipole field during the last 20 million years. However, some individual collections do show local average directions which do not coincide with this world-wide average picture. Most of these anomalies are probably caused by inadequate sampling, uncorrected tectonic tilts, etc., but some appear to represent genuine deviations of a few degrees from the axial geocentric dipole position.

For earlier times it is not possible to test the consistency of observations from igneous or sedimentary rocks on a world-wide scale, as the number of results is fewer, and major tectonic movements have taken place between the different continents (Section 9.3). When it becomes possible to reconstruct the continents unambiguously, then world-wide consistency tests can be

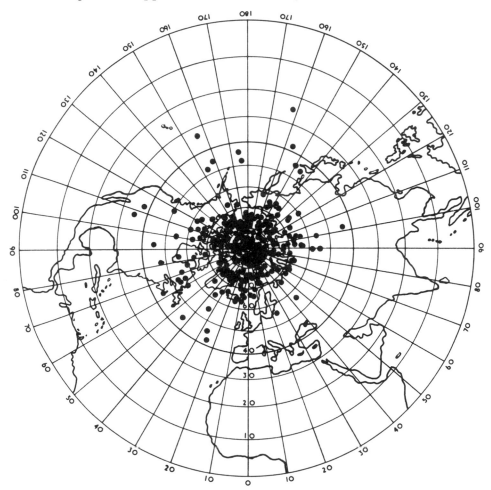

Figure 7.14   Pole positions of igneous rocks up to 20 million years old.

applied, but meanwhile consistency tests are only applicable on a continental or subcontinental scale. For any one period of some 20–50 million years the observed average magnetic declination of sites at different locations on a tectonically stable part of a continent can be plotted (Figure 7.15), thereby defining great circles (meridians) passing through the location of a magnetic pole, irrespective of the magnetic field pattern. Where two circles intersect

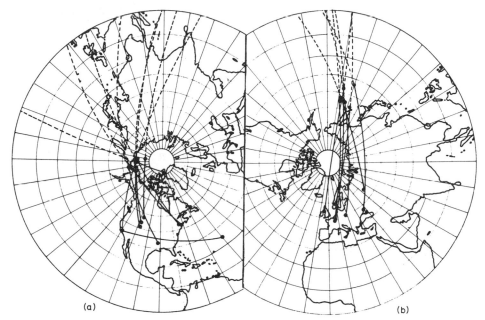

Figure 7.15   Within continent agreement of remanent directions
The pole positions determined using the axial geocentric dipole model for the Earth's past average magnetic field can be compared with those positions which define a common pole by intersection of the palaeomeridians. The poles determined by intersection are identical with those determined using the model, demonstrating the validity of the assumption on a continental scale, during the geological past. Radically divergent directions or poles are explicable in terms of tectonic movements within the continental block (Section 9.4).
(a) Cretaceous rocks of N. America (b) Permian rocks of stable Europe.

lies a pole which is common to the remanent direction of both locations. For any one continent, these great circles intersect each other within a small area of the Earth's surface and poles calculated on the axial geocentric dipole model, using observed values of both inclination and declination, fall in exactly the same area. Only a single geomagnetic dipole can adequately explain these observations within a single continent.

The correspondence between the average geomagnetic dipole and the Earth's axis of rotation is more difficult to determine for these earlier periods. The palaeomagnetic data at present available on the palaeolatitudinal dependence of secular variation, intensity of magnetisation and patterns of directions or poles, etc., are insufficient for any rigorous tests, although they confirm the average dipole nature of past geomagnetic fields. The correspondence must therefore be tested using other indicators of ancient latitudes, such as geological evidence of past deserts, polar ice caps, etc., and the temperature environment of past life forms. These palaeolatitude indicators are unfortunately only coarse measures of latitude since their geographical distributions are substantially affected by local features and patterns of ancient oceanic or wind circulation. Nonetheless, the correlation with these indicators is very good (Section 9.4) and supports the extrapolation of palaeomagnetic evidence for a close association between the average geomagnetic dipole field and the Earth's axis of rotation found in younger rocks to most of the Earth's history.

7.6   Palaeomagnetism and the Earth's interior

The simplest model for the origin of the geomagnetic dipole is an Earth that is uniformly magnetised. This is clearly impossible as it does not account for the changes of the field in both time and space or for its ability to reverse polarity (Section 8.6). Furthermore, the Curie temperature of all known magnetic materials is reached at a depth of only some 20 km and the total intensity of magnetisation of the surface rocks is completely inadequate to explain the observed strength of the field. The only other plausible explanation for the geomagnetic field is generation by electric currents within the metallic liquid core of the Earth. These currents must be maintained by some form of dynamo action as any currents remaining from the formation of the solar system would have decayed long ago and also the palaeomagnetic evidence does not suggest an appreciably stronger field in the past. The circulation of a conducting liquid within the core of the Earth, probably as a result of thermal convection, could produce a dynamo action which would maintain the field, and this would tend to be dipolar as the motion would be influenced by the rotation of the Earth. The actual circulation need only be quite slow, about 4 cm/hour, and is probably extremely complex. The magnetic field in the core would be carried along by the circulating electrically conducting core material, so that the field observed at the Earth's surface, the poloidal field, would be only a small proportion of the total magnetic field of the Earth, and a much larger field, the toroidal field, would be restricted to the core. Reversals of polarity of the poloidal field may therefore reflect only minor changes in the toroidal field.

The present westward drift of the non-dipole field suggests that the mantle is dragging behind the core, but if the magnetic coupling between the core and mantle changes, this drift could either increase or decrease. The palaeomagnetic evidence, at the moment, indicates that there has been a westward drift during most geological periods, but its rate is still unknown. Further detailed studies of the geomagnetic field during a change of polarity (Section 8.5) may clarify this problem, but a critical factor is the difficulty of dating rocks only a few hundred years different in age.

The possible persistence of slow secular change in the Pacific region is particularly important to studies of core-mantle relationships as it suggests some connection between the Earth's surface features and those of the core. This reduced secular variation, if verified, may be explainable by either a suppression of core motions in the Pacific region or by a screening of their magnetic effects by, for example, a conducting layer within the mantle beneath the Pacific. Such explanations suggest that movements of the surface of the Earth are coupled, by some means, with those in the outer core and changes in one area can have direct effects on the other.

Studies of the remanence in samples of extraterrestrial material are still in their infancy. Samples from the Moon and meteorites are difficult to analyse on Earth as their high iron and nickel content means that they readily acquire a viscous magnetisation on entering the Earth's magnetic field. However, preliminary results suggest that the Moon originally had a stronger field than is observed at the moment, and such evidence is clearly of major significance in evaluating models of the evolution of the solar systems.

8

Reversals of magnetisation

8.1   Introduction

The fact that many rocks are magnetised in the opposite direction to that of
the Earth's present magnetic field was one of the earliest significant
discoveries of palaeomagnetism (Section 1.2). This observation raised the
possibility that either the geomagnetic field periodically reverses its polarity
or that the magnetisation in rocks can sometimes be acquired in a direction
opposite to that of the ambient field. Various mechanisms are known by
which the magnetisation of a rock could possibly self-reverse (Section 8.2)
although natural self-reversal is extremely rare. These mechanisms are related
to the presence of specific minerals and it should be possible to decide
whether geomagnetic reversal or self-reversal has taken place by determining
the relationship between the mineral composition and the polarity of the
rock (Section 8.3). Similarly a study of the relationship between the ages of
reversed rocks in different parts of the world also offers a method of
checking the cause of reversed polarity (Section 8.4) and further information
can be obtained from examinations of the way in which the changes of
polarity take place (Section 8.5). Although there are some puzzling features
about the results of these tests, it is certain that reversals of the geomagnetic
field occur and that self-reversal is rare. It is possible, therefore, to construct
a time scale of polarity changes (Section 8.6) which can be used for the
dating of rocks (Section 9.2) and for testing hypotheses concerning the nature
and cause of the geomagnetic field (Section 7.6).

8.2   Self-reversal mechanisms

All self-reversal mechanisms require the material to contain two components
magnetised antiparallel to each other. Reversal then occurs when the
component antiparallel to the original ambient field becomes greater than

114

that of the parallel component. Partial self-reversal can occur when the antiparallel component increases relative to the parallel component but does not become dominant so that the magnetisation of the substance is correspondingly decreased but does not become reversed in direction. All ferrimagnetic and antiferromagnetic materials (Section 2.2) are potentially capable of self-reversal since they contain two antiparallel magnetic lattices which are formed simultaneously as the material becomes magnetised in an external field. It is possible for the antiparallel lattice to increase its magnetisation relative to the parallel lattice by ionic movements, particularly as the material cools (Figure 8.1). Complete self-reversal by this mechanism has been produced in synthetic minerals, and partial self-reversal has been

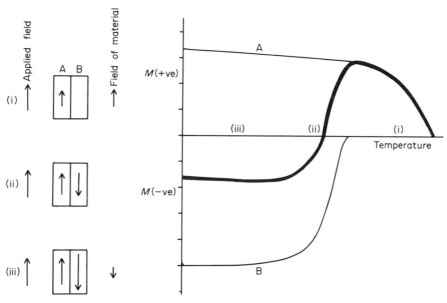

Figure 8.1 Magnetostatic self-reversal

A and B are two substances of different Curie temperature. On cooling, substance A becomes magnetised first (i) parallel to the external field. At a lower temperature, B becomes magnetised in the external field and that of substance A. In this model, substance B is at a critical distance from A so that the field of A is dominant and substance B becomes magnetised anti-parallel to the external field (ii). On further cooling, the remanence of B becomes stronger than that of A and the magnetisation of the two substances together becomes reversed to the applied external field. The increase in remanence of B compared with A may be a natural property of the two substances or may be due to a gradual change by ionic migration, or magnetic instability, in substance A.

found to occur in both magnetite and haematite near their transition temperature (Sections 3.2 and 4.5) so that haematite may show reduced intensity below 100°C. This form of self-reversal could conceivably occur naturally in rocks when Fe^{3+} ions of the parallel lattice are replaced by non-magnetic Al^{3+} ions. However, the amount of aluminium in naturally occurring iron-titanium oxides is normally insufficient for this process to occur, and furthermore, in magnetite, it tends to enter the antiparallel rather than the parallel lattice.

A magnetostatic self-reversal is possible in materials containing two intimately mixed substances of different Curie temperatures (Figure 8.1). As the material cools, the high Curie temperature substance becomes magnetised parallel to the external field but, the second substance becomes magnetised under the influence of both the external field and the back field of the previously magnetised substance. If grains of the two substances are sufficiently close together, it is possible for the second substance to become magnetised antiparallel to the external field. When the total magnetisation of the lower Curie temperature substance exceeds that of the high Curie point substance, the magnetisation of the material self-reverses spontaneously. Similarly, if the high Curie temperature substance is later removed or is less magnetically stable, then the net remanence of the material becomes increasingly dominated by that of the lower Curie temperature, reversely magnetised substance.

Magnetostatic self-reversal has been simulated in very weak external fields using alternating layers of magnetite and pyrrhotite. Pyrrhotite, of particular composition ($Fe_x S$ where $0.8 < x < 1.0$), can spontaneously self-reverse in extremely weak fields if it contains appropriate variations in vacancy distribution within its lattice to give different Curie temperatures in different locations within the same structure. Such self-reversals, however, do not occur in stronger external fields and often require repeated heating and cooling to produce the appropriate interactions between different areas. Magnetostatic self-reversal can also occur as new magnetic materials form from previously magnetised material, e.g. exsolution of pure haematite from a haematite-rich ilmeno-haematite solid solution, or the growth of goethite as an oxidation rim on magnetite. However, it is not certain if magnetostatic or super-exchange coupling can be dominant under such conditions and most secondary minerals appear to become magnetised parallel to the ambient external field rather than to the field of pre-existing minerals.

The only fully substantiated, repeatable self-reversal known in rocks is that of the Haruna dacite in Japan. This dacite is a hypersthene hornblende pumice which was quench cooled on eruption and contains equal amounts of ilmenite and haematite in solid solution (x FeTiO$_3$ .1$-x$ Fe$_2$O$_3$ where $x = 0.5$). The mechanism of self-reversal in this ilmeno-haematite, which has been found in synthetic compositions in the range $0.45 < x < 0.60$, is complex and related to a critical stage of ionic ordering within the solid-solution when there are intimate mixtures of zones of ordered and disordered iron atoms. On cooling while at this critical stage, self-reversing magnetostatic coupling occurs between the ferrimagnetism of ordered areas and the parasitic ferro-magnetism of iron-rich disordered regions with slightly lower Curie temperature. Partial reversal has been observed in some naturally-occurring haematite rich ilmeno-haematite solid solutions, related again to different ionic distributions within the crystal framework, and self-reversal has been produced in these at 100°C and below in the laboratory but only after prolonged heat treatment.

Some self-reversals have also been found in samples of basalt dredged from the ocean floor, but chemical changes on heating mean that the process is not repeatable and is probably related to an interaction between weathered and unweathered minerals. In general, therefore, the conditions for self-reversal are extremely rare in natural rocks.

8.3   Correlations between petrology and polarity

A very strong correlation has been found in several igneous rock sequences between the polarity and the degree of oxidation and exsolution of their iron-titanium oxides. Studies of British rocks 50–60 million years old and Icelandic rocks less than 20 million years old have shown that 80% of samples with exsolved iron-titanium oxides are of reversed polarity and more than 90% of samples with unexsolved iron-titanium oxides are of normal polarity (Figure 8.2). These correlations have also been found in less detailed studies of rocks of widely different ages from various parts of the world. The correlation implies that reversed polarities in rocks are physically linked with the high degree of oxidation and exsolution of their iron-titanium oxides and therefore are due to a self-reversal mechanism rather than a geomagnetic reversal. On the other hand, detailed studies of other igneous rocks have failed to establish any correlation and studies within single lavas and dykes

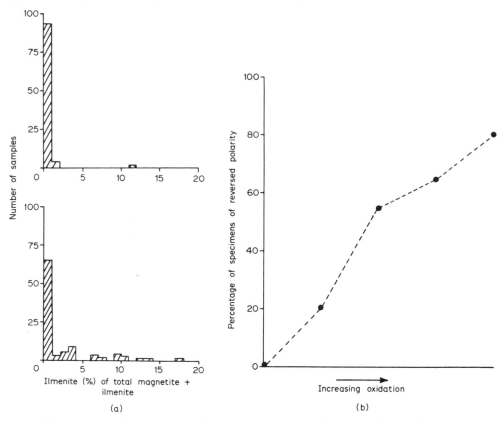

Figure 8.2   The correlation between exsolution of iron-titanium oxides and polarity

Rocks of normal polarity are almost entirely unexsolved; those of reversed polarity generally exsolved. (After Ade-Hall and Wilson, 1969.)

show that the degree of oxidation and exsolution often varies from the margin to the more slowly cooled interior, but the polarity remains the same.

At the moment no physical processes are known by which self-reversal could occur in such rocks. The exsolution and degree of oxidation in many of them must have been established while the temperatures were high, so that their polarity cannot be related to the formation of new magnetic minerals in the back field of previously magnetised magnetic minerals, particularly as many of the highly oxidised minerals are non-magnetic at room temperature. A further puzzling feature arises from the unequivocal evidence that the geomagnetic field periodically changes polarity (Section 8.6) and has been normal and reversed for approximately the same lengths of time. The eruption of highly oxidised rocks should be equally common in different polarity zones and consequently about 50% of any collection of rock samples covering several geomagnetic polarity zones must be of reversed polarity, whether or not self-reversal has occurred.

It is therefore likely that the observed correlations, despite their very high statistical significance, are coincidental unless some extraordinary mechanism can be discovered by which the composition of the mantle of the Earth, which is critical in determining the composition of most igneous rocks, has a variable oxidation state according to the polarity of the geomagnetic field.

8.4   Correlations between polarity and the age of rocks

The first detailed stratigraphical studies of sequences of normal and reversely magnetised rocks were made in Iceland in 1950–51. At that time it was thought that the geomagnetic field reversed regularly, probably every million years, so that variation in the number of lavas within each polarity zone was thought to reflect the rate of eruption. Subsequent improvements in radioactive dating methods, particularly those using potassium/argon, have allowed detailed studies of the dates of polarity zones during the last 4 million years (Figure 8.3). These have shown the occurrence of world-wide major *epochs* of one polarity, lasting approximately 10^6 years, with brief polarity *events* within these epochs when the polarity reverses for some $10^4 - 10^5$ years.

The dates of polarity changes, originally established from igneous rocks erupted during the last few million years, have been confirmed by polarity sequences determined in deep-sea sedimentary piston cores which sample the last 2–3 million years of deposition. The polarity sequences in these cores can be dated from their microfossil content or from the extrapolation of sedimentation rates (although the latter is somewhat imprecise). These observations confirm a world-wide pattern of polarity changes (Figure 8.4), including brief events.

Any correlation between the polarity and age of older rocks is more difficult to establish as the dating is generally less precise than the frequency of polarity changes. During the last 70 million years (Section 8.6), on

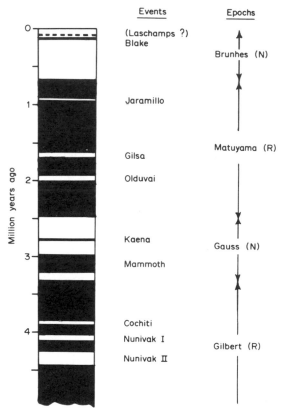

Figure 8.3   The polarity time-scale for the last 4 million years
This scale has been constructed from the polarity and radioactively determined ages of igneous rocks. The Laschamp reversed event has not been fully substantiated and it is possible that other, very brief events, may not yet have been isolated.

present estimates, an average of five changes have taken place during each million years but, as more polarity events are discovered, this frequency could be doubled. In earlier times, however, polarity changes appear to be less frequent and some periods have been determined where the same polarity seems to have been maintained for at least 50–60 million years, e.g. the Kiaman reversed period, which appears to have lasted from some 290–310 million years ago (mid Westphalian/Missourian) until about 230 million years ago (upper Permian). On the other hand at least five polarity changes have been recorded in a 100 million year old grit which is less than

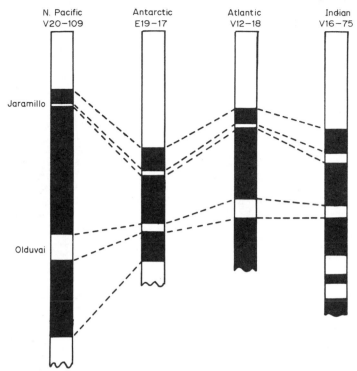

Figure 8.4   The polarity of deep-sea sedimentary cores
Correlation is possible between the polarity sequence in cores from most oceans of the world and has been substantiated by microfossil dating of the polarity changes. The examples shown are taken as representative of a large number of cores in each ocean. (After Ninkovich *et al*, 1966; Hays and Opdyke, 1967; Glass *et al*, 1967; Opdyke and Glass, 1969.)

10 cms thick. It is not possible to test whether these older reversals occurred simultaneously on a world-wide scale as the precision of dating is inadequate, but there is no reason for supposing that the correlation which has been established in younger rocks is not present in older rocks as well.

8.5   Polarity transition zones

The vast majority of stably magnetised rocks are either normally or reversely magnetised, and intermediate zones, in which the polarity changes from one direction to the other, are rare and comparatively brief. The duration of

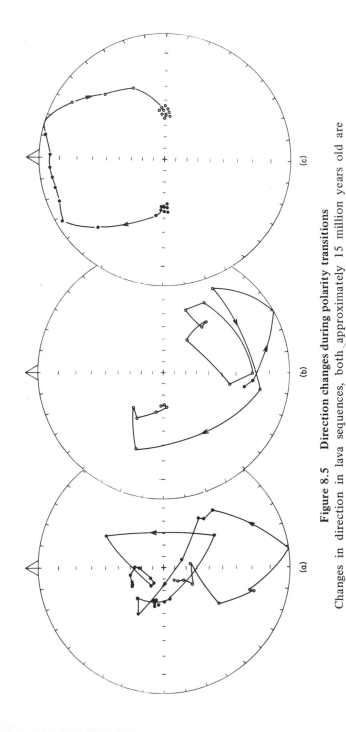

Figure 8.5   Direction changes during polarity transitions

Changes in direction in lava sequences, both approximately 15 million years old are shown in (a) and (b), and in sediments nearly 400 million years old in (c). (After Watkins, 1969; Goldstein *et al*, 1969; and Vlasov and Kovalenko, 1963.)

transition periods cannot be dated precisely although the rate of eruption or deposition of relevant lavas or sediments can sometimes be estimated, and indicates periods of some 5–10,000 years.

A few transitions have been studied in detail, and they are all characterised by a reduced intensity of magnetisation, despite the fact that they are otherwise identical to rocks in adjacent polarity zones. The geomagnetic field appears to have decreased to approximately a fifth of its usual value before the change of direction and to have subsequently recovered after the transition was completed. The actual pattern of directional change is still somewhat uncertain as most reliable studies have been made in lava sequences which do not afford a continuous record. However, both lava and sedimentary sequences so far studied suggest that most reversals take place systematically, the directions changing fairly smoothly along approximate great circles from one polarity to the next (Figure 8.5), although there are usually apparent standstills during the transition. When completed, the reversal is rarely exact, i.e. within 1–2° of 180°, but is more usually some 170–175°. This behaviour is extremely difficult to explain by known self-reversal mechanisms which can only produce an exact reversal, without standstills. The pattern of reversal, particularly the apparent standstills, appears to be related to variations in the strength of the equatorial dipole(s) and non-dipole fields while the axial component is reduced. There also appear to be occasional brief periods of attempted reversal when the axial dipole is reduced in strength and the geomagnetic field direction changes by more than 40°, but, as the axial dipole rebuilds, the direction of the field returns to that of the previous polarity. It appears that when the axial dipole is greatly reduced and the equatorial dipole(s) become the dominant feature of the field, the direction adopted by the re-forming axial dipole depends only on some minor features of the circulation system in the Earth's core (Section 7.6).

8.6   Geomagnetic reversals and the polarity time scale

There can be no doubt that the Earth's magnetic field does periodically change its polarity. This is shown by the world-wide and simultaneous nature of the phenomenon and particularly the correlation of the dates of polarity changes observed in igneous and sedimentary rocks which acquire their remanence by different mechanisms and are dated by different methods. Further confirmation of genuine geomagnetic polarity change is shown by the directional behaviour in transition zones (Section 8.5) and the identify of polarity, in general, between igneous rocks and their baked contacts, both containing radically different minerals. Furthermore, the physical situations in which rocks could self-reverse are geologically rare and only occur under very specialised conditions. This means that the polarity of a rock can almost invariably be attributed to the polarity of the geomagnetic field in which it acquired its remanence and the rock can therefore be dated by establishing its location within a known polarity sequence. Unfortunately the rapidity of reversals means that it is difficult to establish a precisely dated ($\pm 5 \times 10^5$ years) polarity scale for rocks older than a few million years. However, the application of polarity dating to the magnetic anomalies associated with young oceanic igneous rocks has revolutionised concepts of the origin and age of the ocean floors, and has allowed the determination of the sequence of polarity changes during the last 70 million years or so.

Magnetic anomalies associated with the igneous rocks of the ocean floors have been recorded for many decades (the magnetisation of the oceanic sediments is too weak for direct measurement at the ocean surface). Until recently, the main use of these records was to identify the mid oceanic rift system, characterised by a particularly large anomaly. A new development came in the late 1950s when detailed surveys were conducted off the Californian coast and it was found that the oceanic magnetic anomalies form parallel strips. These strips are sometimes offset perpendicularly by inactive faults and the amount of horizontal displacement can be determined by matching the anomaly patterns on either side. The reason for the strip pattern was obscure until detailed surveys were carried out on the Reykjanes Ridge, just south of Iceland. This survey, traversing the mid-Atlantic ridge, showed that parallel strips of positive and negative anomalies also occur in this region, and their pattern is exactly symmetrical about the ridge axis; the anomaly pattern going west from the ridge being the mirror image of that going east. Furthermore, the sequence is exactly the same as the polarity changes of the Earth's field during the last 4 million years (Figure 8.6). The age of the remanence causing the anomalies can therefore be dated by comparison with the established polarity time-scale, and shows a systematic increase away from the ridge centre, corresponding to a spreading rate of

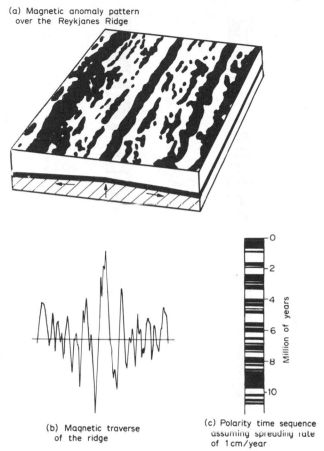

(a) Magnetic anomaly pattern
over the Reykjanes Ridge

(b) Magnetic traverse
of the ridge

(c) Polarity time sequence
assuming spreading rate
of 1 cm/year

Million of years

Figure 8.6   Magnetic anomaly patterns at mid-ocean ridges

Magnetic anomalies observed at the ocean surface show symmetric patterns on either side of the oceanic ridges, illustrated here (a) for the Reykjanes Ridge, just south of Iceland. These anomalies (b) are variations of only 1−2% of the total strength of the Earth's magnetic field but are very distinct. The igneous rocks carrying the remanence causing them can be only some 200−300 m thick and their remanence must have been acquired within a very narrow zone, possibly only 30 m wide, before the rocks separated and were carried out at equal rates in opposite directions. The anomaly pattern therefore provides a magnetic 'tape recording' of polarity changes of the Earth's magnetic field and, if a spreading rate of 1 cm/year is assumed for this region, the sequence and age of polarity changes (c) is identical to that determined from the dating of rocks erupted subaerially (Figure 8.3). (After Heirtzler *et al*, 1966.)

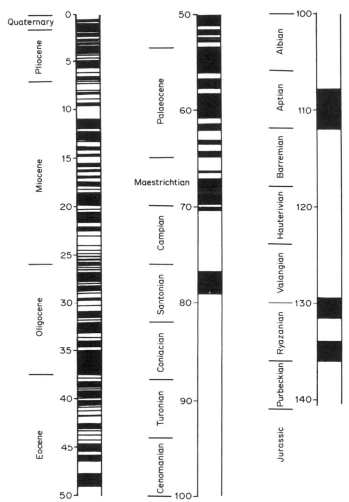

Figure 8.7   The polarity time-scale for the last 140 million years

The polarity sequence can be dated assuming constant spreading rates for the ocean floors and this can then be modified for changes in spreading rates determined from dating of the oldest sediments in different parts of the ocean. For periods earlier than some 70 million years, the polarity sequence can only be dated by conventional geological methods but although these are not precise within 1−2 million years, the frequency of polarity change is less and the pattern is unlikely to be significantly modified if dating techniques are improved. (After Heirtzler *et al*, 1968 and Helsley and Steiner, 1969, modified using preliminary dating of deep-sea sedimentary cores.)

1 cm/year in each direction. Exactly the same sequence of anomalies has now been identified at the central ridges in all oceans, although the spreading rates vary. The anomalies further away from the centre of the ridge can be matched on both sides of the ridge axis and also from one ocean to the next, but they cannot be dated by direct comparison with the established, but short, polarity time-scale. The age of the remanence associated with these older anomalies can, however, be estimated by assuming the same spreading rates as those determined for the last 4 million years and these age estimates can be checked against the age of the oldest sediments overlying the igneous ocean floor using the microfossil content of drill cores. These sedimentary dates have confirmed, within about ±5%, the ages estimated from the spreading rates, so that the anomaly pattern of the ocean floor can be used to construct a time-scale of polarity changes during the last 70 million years. This can be extended further backwards in time using standard stratigraphical dating techniques combined with polarity determinations of continental rocks (Figure 8.7).

8.7   Reversals and evolution rates

The study of the magnetisation of deep-sea sedimentary cores shows that, on a world-wide scale, certain species of marine organisms which are used for dating became extinct and new species of index fossils appeared at times approximately coincidental with polarity changes. The correlation is not exact, although the discrepancies probably arise from the activity of burrowing animals which mix both the fossil remains and the magnetisation in the top few centimetres of oceanic sediments. Some species live through several reversals before dying out, sometimes during an apparently constant polarity zone.

The reason for a correlation between polarity change and speciation is difficult to establish. The simplest explanation is that the reduced strength of the geomagnetic field during a transition (Section 8.5) means that the Earth's surface is no longer protected by the van Allen belts from the Sun's more powerful radiation. However, the effect of radiation on evolution is unlikely to be direct as the magnetic shielding by the ionosphere is less efficient than that by the Earth's atmosphere and is only really effective in Equatorial and tropical regions. Furthermore, the increase in radiation at the

Equator, arising from the loss of the van Allen belts would only be some 10%, and the increased radiation effects would not penetrate more than a few centimetres below the ocean surface. This suggests that the increased speciation may be secondary, possibly occurring in response to climatic changes. Such changes may result from variations in atmospheric circulation produced by changes in the ionisation level of the upper atmosphere. Another possibility is that there is an increase, although very slight, in radioactive C^{14} produced at high altitude and carried by rain to the Earth's surface. It has been suggested that geomagnetic reversals are caused by large scale meteoritic impact, which would have a direct effect on organisms and dramatically increase the dust content in the atmosphere, thereby causing climatic changes and affecting indirectly the speciation of organisms. It is doubtful if the meteoritic impact hypothesis is viable, and an indirect climatic relationship is more realistic, although it is difficult to explain why this should operate on oceanic rather than terrestrial organisms. If, however, increased speciation is related to a reduction in the strength of the geomagnetic field there should be a correlation between species change and the periods of reduced field during attempted reversals (Section 8.5).

9

Geological applications

9.1 Introduction

Palaeomagnetism can be applied to many types of geological problem, most of these relating to either the dating of rocks (Section 9.2), which often throws light on their mode of origin, or to estimates of the amount of tectonic movement which rocks have undergone since they acquired their magnetisation (Section 9.3). When studying these problems, the same criteria must be applied to the selection of palaeomagnetic data as are applied in geomagnetic studies (Section 7.1), so that the isolated primary remanent magnetisation can be considered to be an accurate reflection of the original geomagnetic field. However, there are yet more assumptions involved in the use of palaeomagnetism in geological problems, arising from the need to adopt a model for the previous geomagnetic field and its secular variation in order to compare observations from different areas. The usual model assumes that the average field is that of a dipole (Section 7.5) and, in some applications, that this dipole is geocentric and coincident with the Earth's axis of rotation. In order to obtain this average field, it is necessary to average out the effects of secular variations (Section 7.3) and this limits the statistical accuracy with which the average direction or pole can be determined (Figure 9.1).

9.2 The dating of rocks

Using palaeomagnetic methods, rocks can be dated at three levels of precision; by secular variation cycles with periodicities of some 10^3 years, by use of polarity zones and polarity transitions with periodicities of $10^3 - 10^7$ years and by determinations of average geomagnetic pole positions showing variations of some $10^6 - 10^9$ years.

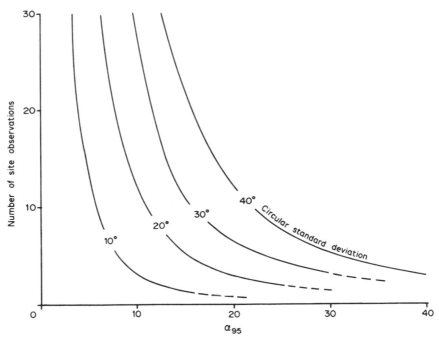

Figure 9.1   Secular variation and palaeomagnetic precision
Secular variations give rise to an intrinsic scatter of directions of remanence and pole positions which can be defined in terms of the circular standard deviation. This intrinsic scatter limits the precision, α_{95}, with which the mean can be determined, for any number of spot readings (N) which systematically sample the variation cycle.

(a)   Few secular variation cycles have been studied in detail (Section 7.3) but it seems probable that each cycle has a characteristic pattern which, if identified in two or more localities, can be used for precise correlations on time scales as short as $10^2 - 10^3$ years. These correlations involve detailed analyses in each location and should therefore only be considered in areas where there is a strong probability that the rocks are of comparable age and the palaeomagnetic properties are adequate for the analyses to be completed in both areas. Such correlations have been carried out successfully between different exposures of the same lava sequence in order to determine the amount of vertical displacement on a fault between the two exposures. It is unlikely that such precise correlation can be obtained in sediments where the process of magnetisation is long and comparable to the duration of a secular variation cycle (Section 4.2).

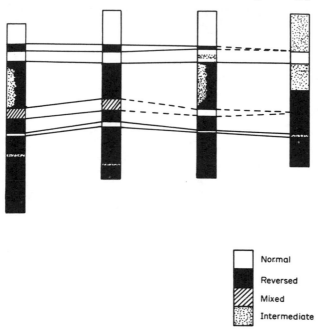

Figure 9.2   Correlation using polarity of remanence
Sections of the Triassic Chugwater Formation (red sediments some 200 million years old) separated by several 10s of kilometres can be correlated by the pattern of polarity changes although this must be supplemented by petrological correlations because of the difficulty in obtaining a continuous sequence of stably magnetised rock. (After Picard, 1964.)

(b)   The irregularity of geomagnetic reversals (Section 8.6) makes it possible to match polarity sequences in rocks from one exposure to the next, particularly if the rate of deposition or eruption was approximately constant in both areas. In most cases the irregularity of rock accumulation tends to obscure the pattern of the polarity periods and precise matching is difficult over extensive areas, unless supplementary correlations (e.g. stratigraphic marker horizons) can be used (Figure 9.2). However, such horizons are often diachronous compared with the almost instantaneous occurrence of polarity transitions.

When the polarity sequence has been dated (Section 8.6) the polarity time scale can be used for absolute dating, as in the case of the remanence of oceanic igneous rocks (Figure 9.3). Unfortunately the time-scale of polarity

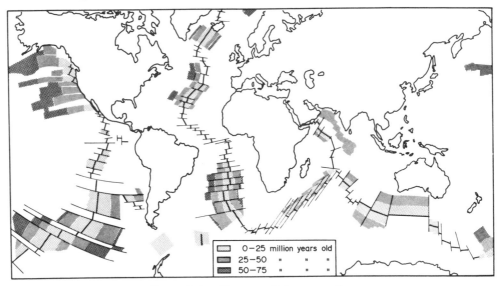

Figure 9.3   The age of the ocean floor
The remanence of the igneous ocean floor can be dated using the polarity time scale
(Figure 8.7) mainly determined from the polarity pattern in the South Atlantic Ocean.
(Modified from Heirtzler, 1968.)

changes for rocks older than about 150 million years is less well established,
although it is possible to distinguish, for example, sequences of mainly
Triassic (190–230 million years) rocks of both polarities from sequences of
lower-middle Permian (230–280 million years) rocks of entirely reversed
polarity. Such applications will become more widespread as knowledge of
the polarity time-scale increases.

As each separate polarity transition is probably characterised by a specific
pattern of field changes, particularly by the number and duration of
temporary stand-stills (Section 8.5), it seems probable that it will ultimately
become possible to correlate polarity changes on a fine scale by comparison
of detailed analyses of the transition zones between polarities.

(c)   Rocks of increasing age from the same continent yield average
palaeomagnetic pole positions which are increasingly displaced from the
Earth's present axis of rotation (Section 7.5). It is therefore possible to
construct a curve connecting the average geomagnetic pole positions for each
continent during geological time. Such *polar wandering curves* have been

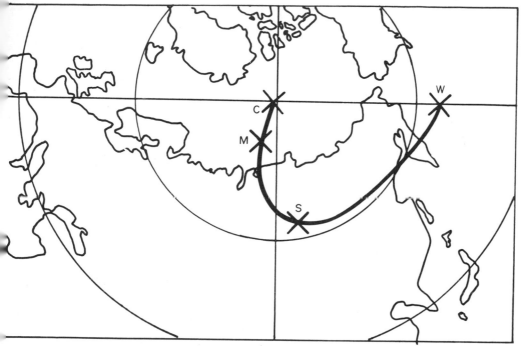

Figure 9.4   Dating from known positions of palaeomagnetic poles
The pole positions for rocks of unknown age in Czechoslovakia are shown in relation to
known Czechoslovakian pole positions and can be dated as (C) less than 20×10^6 years,
(M) $30-50 \times 10^6$ years (S) $50-65 \times 10^6$ years and (W) $155-175 \times 10^6$ years. (After
Hanus and Krs, 1963.)

determined with varying precision for most continents for the last 300
million years (Figures 9.4 and 9.6) and these curves are gradually becoming
better defined and being extended to earlier periods. It is therefore possible
to determine the age of remanence of a suite of rocks by determining the
position of its average palaeomagnetic pole on the known polar wandering
curve for the same tectonically stable block (Figure 9.4). The precision of
this dating naturally depends on the precision of both the established polar
wandering curve and the new observations, and also on the rate at which the
pole is moving along the curve at that time. This movement is, on average,
some 0.3° per million years, but the pole has sometimes remained
approximately constant in position relative to individual continents for up to

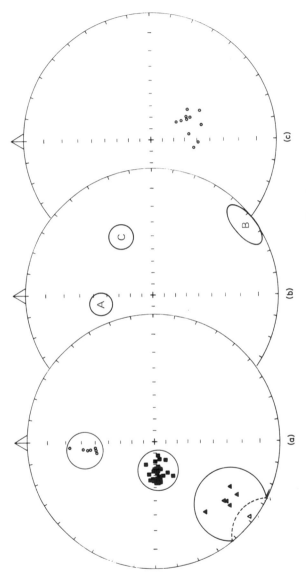

Figure 9.5   Dating by remanent directions

(a) Dykes of similar orientation in Canada, the Abitibi Swarm, show three distinct groups of directions and are therefore probably of three different ages. (b) Basic igneous rocks from Colorado show three main groups of directions corresponding to ages of (A) 60 x 10^6 years, (B) 1300–1400 x 10^6 years and (C) 450–600 x 10^6 years. (c) Directions of remanence from 8 separate igneous plugs, the Monteregian Hills, near Montreal, show identical directions and have subsequently been shown to all have similar ages. (After Larochelle, 1961, 1966 and Mutschler & Larson, 1969.)

$10^7 - 10^8$ years so that a precision of some 10^6 years can only be obtained during rapid relative movement (Section 9.3).

Differences in pole positions can be significant in determining the origin of certain types of rock even when the polar wandering curve has not been established (Figure 9.5) since such differences imply differences in age, even though the magnitude of such difference cannot be assessed. Iron ores, for example, which are often difficult to date by conventional methods, may form directly by deposition or igneous activity (primary ores), by metamorphic activity associated with later igneous activity (secondary ores) or by concentration during the weathering of pre-existing rocks (residual ores). Primary ores are of similar age to the surrounding rocks and therefore both have similar directions of remanence, secondary ores have directions of remanence similar to those of the igneous source, and residual ores have directions of remanence which were acquired later and therefore differ from those of other rocks in the region.

The variation of other properties, such as intensity of remanence, can also be used to indicate age differences, distinguishing, for example, between

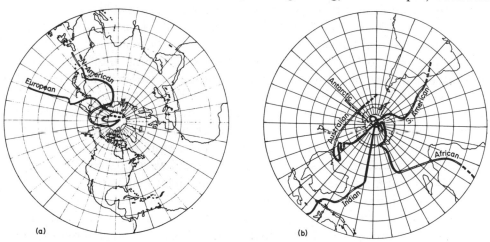

Figure 9.6   Polar wandering curves
The curves are shown marked in 100 million year divisions for (a) northern continents and (b) southern continents, increasing in age away from the present geographical poles. Most details of these curves have *not* yet been defined; the accuracy of the solid curves are probably within some $5-10°$, while the broken curves are more poorly defined. (After Creer, 1970.)

extrusive and intrusive rocks, or dividing massive deposits into their component parts. However, these applications are more closely related to the geological history of the rocks (Section 9.5).

9.3   Tectonic and structural applications

The magnetic anomaly strips of the igneous ocean floors mark growth lines which can be dated by reversals of the geomagnetic field (Section 8.6). These age studies show that most of the ocean floors have formed during the last 150 million years and during this time they have been growing and spreading symmetrically away from the oceanic ridges (Figures 8.6 and 9.3). This age distribution, combined with seismological studies, has led to the recognition of vast *plates* composing the surface of the Earth. These plates are some 60–120 km thick and comprise oceanic and continental crust with some upper mantle material, i.e. the *lithosphere*. Each individual plate behaves as a stable block, and present tectonic activity is concentrated along its margin. If the positions of adjacent plates are adjusted so that anomalies of the same age match each other at the centre of the ocean, it is possible to reconstruct the past geometrical arrangements of the continents. In areas of ocean floor where an allowance can be made for the magnetisation induced in the rocks by the present Earth's field, it is possible to determine the inclination of natural remanence and thus the palaeolatitude at which that part of the igneous ocean floor acquired its magnetisation. Unfortunately this determination is difficult to evaluate as the observed remanence includes various viscous components. Nonetheless the matching of the anomaly patterns allows unique reconstructions to be made for various times during the past 70 million years when reversals of the Earth's field were sufficiently frequent for well-defined and dated anomalies to be distinguished and correlated, but this becomes more difficult for the 70–150 million year old ocean floor which has very few anomalies. Previous polarity changes are variable in frequency and still uncertain but little ocean floor of this age remains, most of it having been absorbed into the Earth's mantle. (The process of present-day absorption is clearly marked by the occurrence of earthquakes within slabs of lithosphere descending beneath the oceanic trenches.)

For earlier times, and to confirm the later sea-floor spreading reconstructions, the past distribution of the continents can only be determined by conventional geological and palaeomagnetic techniques. Geological methods consist of matching structures, rock types, fossils, etc., from one continent to another. This matching is difficult to assess objectively and becomes increasingly ambiguous in older rocks. However, it is possible to define the spatial relationship of different tectonic blocks of any particular age by palaeomagnetic methods and the precision of such reconstructions is dependent only on the accuracy with which the average geomagnetic pole can be defined.

The average declination and inclination of stable remanence in a suite of similarly aged rocks defines the location of an average geomagnetic pole. As this pole can be assumed to be the rotational pole (Section 7.5), the previous latitude and orientation of the sampled area, calculated with respect to this palaeo-pole, can be extrapolated to other parts of the same tectonically stable block and thus the orientation and latitudes of the entire block can be defined relative to the ancient pole (Figure 9.7). When the pole is defined for two separate blocks for the same time the relative latitudes and orientation of both blocks can be determined, although their absolute longitudinal relationship cannot be defined. However, their complete spatial relationship can be defined if sufficient palaeomagnetic poles of different age can be determined for polar wandering curves to be defined for each tectonic block. For periods when these polar wandering curves (Figure 9.6) are identical in shape, the blocks can be moved so that their polar wandering curves coincide, thereby precisely determining the relationship between the two blocks. Differential movement can then be dated from the time that the polar wandering curves separate (Figure 9.8).

At present, the polar wandering curves for different continents are rather poorly defined. This is because much of the available data are untested for stability of remanence and, after adequate criteria have been applied to remove the data from rocks likely to have been affected by remagnetisation, tectonic movements, secular variations, statistical fluctuations, etc., insufficient data remains for curves to be accurately defined. Nonetheless, palaeomagnetic studies of rocks less than 300 million years old have confirmed, in general, previous geological concepts of continental asemblages and palaeolatitude positions. For earlier times, the polar wandering curves are even more poorly defined; many of the present inconsistencies arising from the uncertainties of the age of the rocks and their remanence rather

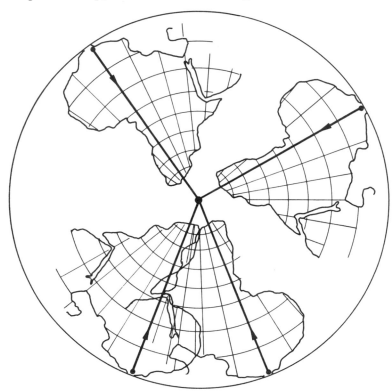

Figure 9.7   Palaeolatitudes and orientation from a single pole position
The palaeomagnetic data for sites of the same age can be used to define the palaeolatitudes over a wide area and also the orientation of the area relative to the pole. However, the data, on its own, can be satisfied by any longitudinal position of the continent relative to the ancient pole.

than from the unreliability of the palaeomagnetic determinations. Clearly this situation is improving rapidly and studies of rocks of all geological ages have demonstrated that there have been movements of different tectonic blocks relative to the average pole at all times. It is, however, important for geotectonic studies and theories of the interior of the Earth, to determine whether such polar wandering curves result from plate motions alone or if there is an additional movement of the entire surface of the Earth relative to the pole, which must remain fixed in space relative to the ecliptic.

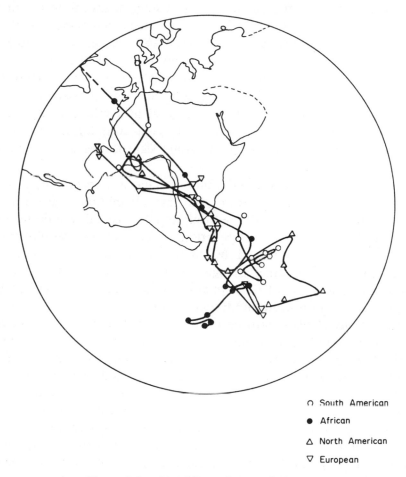

O South American

● African

△ North American

▽ European

Figure 9.8   Matching polar wandering curves

When the polar wandering curves become better defined, precise matching of all
continents at different times will become possible. Meanwhile it is more realistic to
compare proposed matches of the continents such as the geometric fit, against the
corresponding match of their polar wandering curves. In general, the present discrepancies
in the match reflect times for which the pole position has only been estimated or its
location or age is still poorly defined. The poles shown are average positions determined
for different times during the last 500 million years, decreasing in age from the top of the
diagram.

The mechanics of plate movements are not known in detail, but the motion appears to be part of a convective system operating in the mantle. This system is driven by lateral variations in density and temperature, with ascending mantle material at the mid-oceanic ridges and descending lithosphere at the oceanic trenches. A summation over the surface of the Earth of all plate motions must reflect the distribution and magnitude of these vertical parts of the convective system and any persistent, common motion will represent the motion of the total lithosphere relative to the interior of the Earth. Any component of 'polar wandering' due to this common motion is of no direct geotectonic significance as it is purely a consequence of plate movements.

It is unlikely that the observed polar wandering curves, which do not differ in their general nature during the last 70 million years from that during earlier times, can be attributed only to a 'statistical' wandering of the lithosphere relative to the axis of rotation as there are several observations that cannot be readily explained by plate movements alone. Most polar wandering curves show distinct changes in direction within at least the last 50 million years and some of these changes, although poorly defined at the moment, appear to have been simultaneous. Such common movement could reflect interactions of plate motions in response to changes in the convective pattern, such as would result from part of the system being overidden by a continent. However, a 'major discrepancy arises from comparisons of plate movements, as defined from sea-floor spreading, and the motion of the continents relative to the pole, as determined by palaeomagnetic observations. For example, oceanic anomalies and transform faults (Figure 9.3) indicate that the North Atlantic has been opening as a result of sea-floor spreading in a northwest-southeast direction at a rate of some 2 cm/year for the past 55 million years. In contrast, palaeomagnetic data from Greenland and Europe indicate an essentially northward movement of Greenland at 5.4 cm/year and of Europe at 2.6 cm/year during the same period. This implies that the observed sea-floor spreading is superimposed on a net northerly motion of some 2–3 cm/year, which is common to both plates and carries the sea-floor spreading mechanism with it.

It has been suggested that a loading on the lithosphere, such as the creation of the Himalayas or mass transfer at the top of the upper mantle, could cause a change in the Earth's moment of inertia and that this could result in an adjustment of the entire lithosphere relative to the Earth's axis of rotation. Such an adjustment would result in the operation of a force on the lithosphere plates in addition to that associated with convection. However, any such force on the lithosphere alone would enhance plate movements parallel to it while retarding antiparallel movements and therefore would result in asymmetric spreading away from oceanic ridges where they are aligned perpendicular to the force. No significant asymmetry of spreading has been reported on the sinuous oceanic ridges formed during the last 70 million years (the asymmetry south of Australia is small) although spreading rates and directions have varied. This indicates that, apart from convection, any other force on the lithosphere alone is extremely small and geotectonically insignificant. Therefore any significant overall motion, not attributable to plate movements, must affect the entire convective system, including the lithosphere, and most probably reflects either differential movement of the mantle and crust relative to the core or a movement of the entire Earth relative to the axis of rotation.

Various hypotheses have been put forward suggesting that the Earth has been expanding since its formation some 4.6×10^9 years ago, and it has been suggested that continental drift may be a result of such expansion. Most of these hypotheses involve either thermal expansion or changes in the gravitational constant resulting in increases in the volume and surface area of the Earth during geological time. As an increase of 1% in the Earth's radius requires an increase in the surface area of 2%, any expansion of the Earth must involve the addition of new material to the Earth's surface. The world's oceans are generally less than 150 million years old and it is conceivable that they formed as a result of Earth expansion during this time. Such a rate of expansion does not seem feasible by any known physical mechanism, but it is difficult to reject the hypothesis on this ground alone. Fortunately it is possible to use palaeomagnetic observations to determine the Earth's radius at different times. If geomagnetic pole positions are determined for two widely separated regions on the same tectonically stable block, these poles should correspond if the radius of the Earth at the time of the formation of the rocks was the same as it is today. However, if any one of the requirements for tectonic stability, averaged secular variation, constant radius, etc., is not satisfied, then the pole positions would not be identical (Figure 9.9). Unfortunately there are very few parts of the Earth's

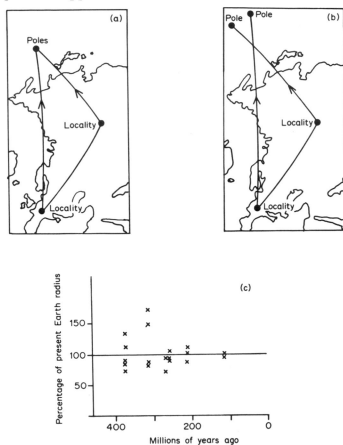

Figure 9.9   The ancient radius of the Earth
(a) If the radius of the Earth has remained constant, then the pole position for rocks of the same age should be identical in present day coordinates for separate localities on the same tectonically stable block. However, if the radius of the Earth has changed (b), then the pole position will no longer correspond but will diverge by an amount proportional to the increase or decrease in radius. (c) Palaeomagnetic determinations of the previous radius are variable, but a lot of the differences are attributable to imprecision of the pole determinations, or to the occurrence of some tectonic movement between the locations since they acquired their remanence. Nonetheless, it is clear that the data is adequate to rule out any extremely rapid increase or decrease in the size of the Earth during the last 300 million years. (Data after Van Andel and Hospers, 1968.)

crust of large enough dimensions which have been stable for prolonged times, particularly for periods longer than the last 300 million years. However, examination of the palaeomagnetic evidence from two such areas, Eurasia and North America, suggests no significant change in the Earth's radius during the last 300 million years. The data for older rocks are inconsistent (Figure 9.9c) some indicating a somewhat smaller radius, some suggesting a slightly larger radius, but in most cases neither the palaeomagnetic nor the tectonic stability requirements for these crustal blocks are fully satisfied. The palaeomagnetic evidence is therefore inconsistent with any major change in the radius of the Earth although a very small increase or decrease over geological time cannot be ruled out on the present evidence. Since continental drift as a result of Earth expansion requires some two-thirds of the Earth's surface to have formed during only the last 150 million years, and the palaeomagnetic evidence shows that this rate of expansion is not possible, this cannot be considered a viable mechanism for explaining continental drifting during this period.

Palaeomagnetism can of course be used to study the movements between different tectonic blocks within the same continent and, as in the case of continental displacements, it is easier to detect rotational or latitudinal displacements than longitudinal movements unless sufficiently accurate polar wandering curves can be established for both blocks. There are many examples of studies of undeformed but rotated blocks, particularly in the western Mediterranean region (Figure 9.10) where rotations occurred during the formation of the Alps and Pyrenees. The same palaeomagnetic techniques can be applied to many other types of geological structure (Figures 9.10 and 9.11). There is clear evidence that some mountain chains have been bent (oroclines) since their original formation, e.g. the Rocky mountains of the western U.S.A. curve round the Mendocino arc, while others, such as the Appalachians of eastern U.S.A. seem to have formed originally as arcuate structures (although the palaeomagnetic evidence for this is still imprecise and therefore somewhat conflicting). Similarly island arcs may apparently form as arcuate structures or they may be distorted subsequent to their initiation. Considerably more care is required in such regional studies if samples are taken from actively deformed areas, as these are likely to have marked secondary magnetisations, particularly in highly deformed rocks e.g. nappe structures (Figure 9.11). However, the unravelling

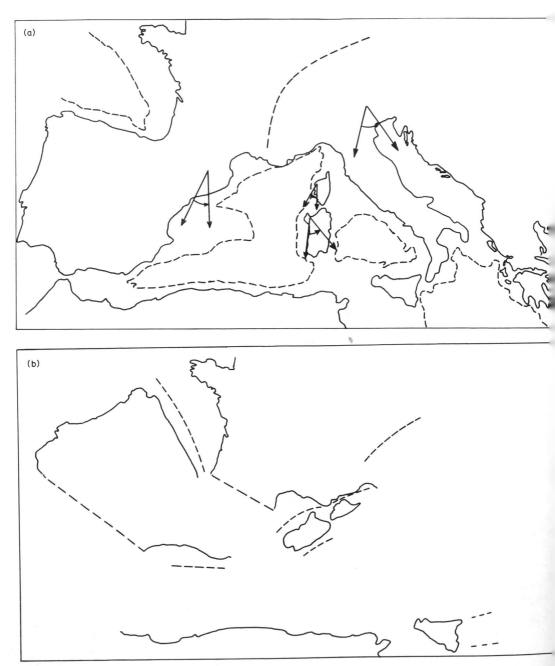

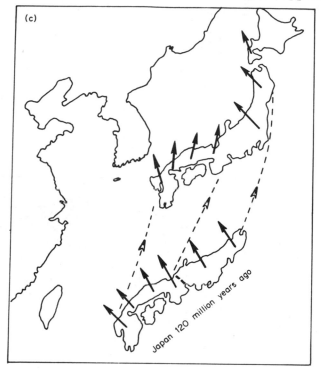

Figure 9.10   Tectonic applications in the Mediterranean and Japanese areas
The movement of different tectonic blocks within the Mediterranean region is becoming
clarified, but there is generally insufficient data to determine the precise age of these
rotations, although most of them appear to be during the last 50 million years. The
palaeomagnetic evidence from the Japanese islands shows that the islands have been
twisted into their present arcuate form during the last 120 million years and that the
islands also lay further south than their present location. It is not known, however, if
their northward movement since then has formed as part of regional movement of the
Asian mainland or has been independent of the tectonics of Asia. (After Kawai *et al*,
1969; Sasajima *et al*, 1968; Zijderveld *et al*, 1969, 1970; Nairn and Westphal, 1968.)

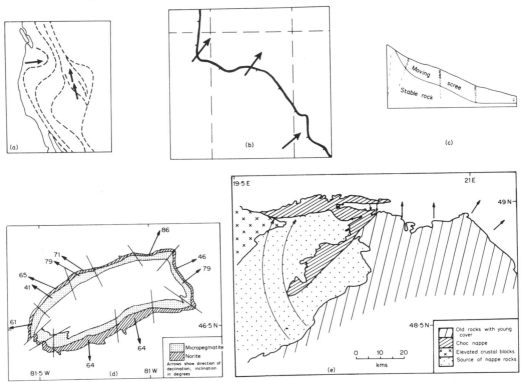

Figure 9.11   Illustrations of some tectonic applications
(a) Evidence from the western U.S.A. suggests that the Mendocino twist of the Rocky Mountain system was formed after the main structural trends had been established (after Irving, 1964). (b) Remanent directions of the same age along the Lewis Thrust have been unaffected by relative movements suggesting that the bend in the outcrop of the thrust is primary in origin (after Norris and Black, 1961). (c) The changes in inclination of remanence in samples taken from drill cores in Tasmania, Australia, allowed the level at which scree movement was taking place to be determined so that engineering work could be sited within stable rock (after McDougall and Green, 1958). (d) The directions of remanence in different outcrops of the Canadian Sudbury nickel deposits (inclination shown numerically) allow the general structure and tectonic evolution of the deposit to be determined allowing more efficient planning of the exploitation of the deposits (after Larochelle, 1969). (e) Directions of remanence in different parts of the Choc nappe in Czechoslovakia allows the evolution of the structure to be determined; arrows showing the probable direction of movement of the rocks (after Kotasek and Krs, 1965).

of the palaeomagnetic history of such rock samples can yield considerable information on the age and thermal evolution of such regions (Section 9.5).

On a more local scale, rock magnetic techniques can be used to supplement standard geological methods for analysing the fabric of rocks which is of importance in determining the conditions of formation and subsequent tectonic history of rocks. In sedimentary rocks, elongate particles are often aligned in the direction of current flow and, in igneous rocks, similar mineral alignments may arise from flow while the magma is still liquid, e.g. foliation in granites and flow banding in lavas, or by gravitational settling of dense minerals, as in some basic rocks. Rocks which have been subjected to tectonic stress usually show secondary alignment of minerals resulting from either physical rotation or preferential growth in directions perpendicular to the stress. The fabric of rocks is usually investigated microscopically by measuring the orientation of particular minerals, e.g. quartz grains. These measurements are tedious and it is possible to determine rapidly the general nature of the fabric by measuring the magnetic anisotropy of a specimen, for although six separate measurements are required to define the ellipse of anisotropy, these measurements can be made quickly using standard palaeomagnetic instruments (Section 5.4), and even finer analyses can be attempted if the number of measurements are increased, although high sensitivity ($\sim 10^{-8}$ gauss oersted^{-1}) instruments are required for this. It is particularly convenient, from this point of view, that the common magnetic minerals, magnetite, haematite and pyrrhotite, are all stress-sensitive and so the magnetic fabric of a rock is sometimes a more sensitive indicator of previous stresses than many non-magnetic minerals.

9.4 Palaeomagnetism and palaeolatitudes

The close association of the average geomagnetic pole and the rotational pole can be demonstrated for the last 25 million years and there are strong theoretical grounds for considering that this existed at earlier times (Sections 7.5 and 7.6). Palaeomagnetically determined past geomagnetic latitudes (Figure 9.7) should therefore correspond with past geographical palaeolatitudes and should correlate with other latitude-dependent features of the same period. There are two main kinds of palaeolatitude indicators; rock types which can only form in quantity under certain climatically controlled conditions, and fossils which indicate the nature of the environment in which they lived. These indicators are only coarse measures of ancient latitude for although climate is largely latitude-dependent the conditions for their formation and preservation were functions of the total environment and this can be strongly influenced by the distribution of land and sea and the circulation of winds and oceanic currents. Furthermore, it is difficult to extrapolate the present condensed latitude zonation of climate into past geological times and many of the past species of flora and fauna were different to those existing today. It is doubtful, therefore, whether most indicators can define a past latitude position within some $20°$, which is considerably less precise than that of the geomagnetic pole and geomagnetic latitude determinations.

Nevertheless for low latitudes, the agreement between the occurrence of latitude indicators and palaeomagnetic determinations is extremely good (Figure 9.12). The characteristics of high latitudes are more difficult to establish, there being a paucity of flora and fauna and lack of low latitude deposits, rather than a presence of specific indicators, other than glacial deposits. However, mountain glaciation can occur in equatorial areas, as in New Guinea and East Africa today, and so indications of polar ice sheet glaciation require the distinction of several features over a large area, e.g. tillites, varves, glacial striae etc. There are spectacular examples of polar ice-sheet glaciation having once occurred in areas which are now in low latitudes, e.g. the Sahara, (450×10^6 years ago) Antarctica, India, Australia, South Africa, and South America ($250\text{-}350 \times 10^6$ years ago), and palaeomagnetic observations are in agreement with the high palaeolatitudes implied by widespread glaciation of all these are as at those times (Figure 9.8).

The determination of palaeolatitudes by palaeomagnetic means has several practical applications which can be of economic importance. For example, the palaeolatitude environment of a variety of animals and plants has already been established (Figure 9.12) and that of the coal flora and fauna is particularly interesting. Most coal deposits formed before 300 million years ago had a polar distribution whereas most later coals had an equatorial distribution, both types occurring simultaneously for some $50\text{--}100$ million years. Again, most large oil deposits are known to have formed within $30°$ of the equator (Figure 9.12) so that the palaeolatitude of the source rocks is of

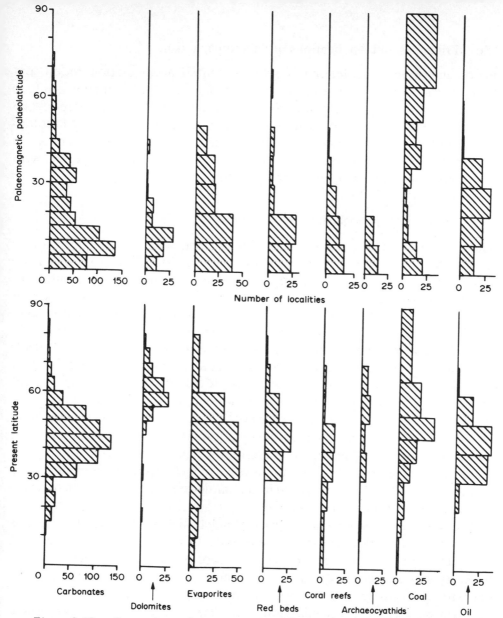

Figure 9.12   Comparison of the present location and palaeolatitude location of various rocks and fossils

The data are only those from which there is some palaeomagnetic control and where substantial remains can be located. (After Briden and Irving, 1964; Irving and Gaskell, 1964; Deutsch, 1965.)

149

major importance in determining the potential accumulation of oil and natural gas. In the same way, the occurrence of certain minerals such as gypsum, anhydrite, laterite, etc., is only possible if the areas were at one time in the appropriate low palaeolatitude. Such applications of palaeolatitude studies are only just commencing, and may soon be extended to determine the distribution of other life forms in the past, together with many other features, such as the orientation of past stress fields.

9.5   The geological history of rocks

The identification of the iron-oxide minerals and their Curie temperature is frequently undertaken as an integral part of standard palaeomagnetic investigations, and this information can be directly related to the conditions of formation of the rock. In igneous rocks, for example, the presence of intermediate composition ilmenite-haematite or magnetite-ulvospinel solid solutions (Section 3.2) indicates that the rock must have cooled extremely rapidly. However, the titanium content of the minerals, which also controls the Curie temperature (Section 3.4), is partially temperature dependent so that the Curie temperature can be used to some extent as a geological thermometer. In metamorphic rocks, for example, the TiO_2 content varies from less than 1% in greenschist facies ($<400°C$.) to over 2% in the granulite facies ($>550°C$.). The temperature of formation of tuffs can also be determined by studying the uniformity of remanence. Tuffs which were deposited hot have a uniform thermoremanence, while tuffs which were erupted hot and cooled before their accumulation can only carry a uniform but weak deposition remanence.

The intensity of remanence is usually much higher at the chilled margins of igneous rocks, although the susceptibility is usually uniform throughout the rock. The identification of rapidly cooled margins by means of the intensity/susceptibility (Q) ratio can therefore be used to divide massive deposits, such as ignimbrite sheets, into separate eruptions. This property can also be used to distinguish intrusive igneous rocks which have slowly cooled margins from extruded rocks with a rapidly cooled upper margin. The observed high intensity of quench-cooled rocks is particularly relevant to the study of the magnetisation of oceanic rocks and the mechanism of emplacement of new igneous rocks in the rift valleys of the mid-oceanic ridges. The new oceanic rocks are quench-cooled on penetrating wet sediments or when coming into direct contact with sea water and are therefore characterised by high Q ratios (Q≈20). The magnetic anomalies over the oceans, which are mostly carried by the rocks in the top 200—300 m of the igneous floor, must therefore be almost entirely related to rocks which have either been in contact with the water itself or with wet sediments, and are not therefore associated with the dykes feeding them. The decrease in intensity of these rocks away from the rift, i.e. in the region where they are more than 2—3 million years old, can be explained by spontaneous viscous decay of this high intensity initial remanence, although fine grained material of high stability retains a significant proportion of the remanence. This means that the possible mechanisms by which the ocean floor is growing by the addition of new material from the mantle are severely restricted, particularly as this stable remanence must be acquired in a very narrow zone in order to account for the sharpness of the anomalies and the detection at the ocean surface of very brief polarity events recorded in the igneous rocks beneath.

The direction and amount of uniaxial pressure exerted on rocks can obviously be determined from the degree of anisotropy which has been imposed, and the amount of recrystallisation can be used as a measure of the amount and possibly the duration of any temperature rise during the history of the rock. In some cases, the direction of remanence can be used to date the temperature decrease at the end of a thermal event. Similar studies have allowed determination of the heating effects of (i) massive intrusions, e.g. the remanence acquired by the Beacon Sandstones of Antarctica from the overlying Ferrar dolerites, (ii) local heating during burial, e.g. the Old Red Sandstone of the Anglo-Welsh Cuvette, and (iii) the regional heating accompanying phases of mountain building, e.g. the Bloomsberg beds of the U.S.A. Similarly, the depth of ancient weathering can be traced by its remagnetisation effects. The presence of maghemite ($\gamma Fe_4 O_4$) is also diagnostic of weathering as it can only form naturally by oxidation at low pressures, i.e. near the surface.

Selected bibliography

General Works

Lists of palaeomagnetic directions and pole positions appear periodically in *Geophys. J.*, under various authorships, giving compilations and references to recently published data.

Blackett, P. M. S. "Lectures on rock magnetism," Weizmann Science Press, Jerusalem, p. 131 (1956).

Collinson, D. W., K. M. Creer and S. K. Runcorn (Eds.) Methods in Palaeomagnetism, "Developments in Solid Earth Geophysics" 3, Elsevier, Amsterdam, p. 609 (1967).

Cox, A. and R. Doell. Review of Paleomagnetism, *Bull, Geol. Soc. Amer.,* **71**, 645-768 (1960).

Doell, R. R. and A. Cox. Paleomagnetism, *Adv. Geophys.,* **8**, 221-313 (1961).

Irving, E. "Paleomagnetism and its application to geological and geophysical problems," J. Wiley and Sons, New York, p. 399 (1964).

Nagata, T. "Rock Magnetism" (2nd Edition) Maruzen, Tokyo, p. 350 (1961).

Runcorn, S. K. (Ed.) "Palaeogeophysics," Academic Press, London, p. 518 (1970).

Strangway, D. W. "History of the Earth's magnetic field," McGraw-Hill, New York, p. 168 (1970).

Wilson, R. L. Palaeomagnetism and rock magnetism, *Earth-Sci. Rev.,* **1**, 175-212 (1966).

Historical (Chapter 1)

Chevallier, R. L'aimantation des laves de l'Etna et l'orientation du champ terrestre en Sicile du XIIe au XVIIe siecle, *Ann. Phys.,* **4**, 5-162, (1925).

Gilbert, W. "De Magnete," Dover, New York, p. 368 (1958). (English translation of the original 1600 publication by P. F. Motteley).

Needham, J. "Science and Civilization in China," Vol. 4, Cambridge University Press, Cambridge, p. 434 (1962).

Smith, P. J. Petrus Peregrinus' Epistola, *Earth Science Revs./Atlas,* **6**, A11-17 (1970).

Physical Basis (Chapter 2)

Brailsford, F. "Physical principles of magnetism," Van Nostrand, London, p. 274 (1966).

Chikazumi, S. "Physics of Magnetism," Wiley and Sons, New York, p. 554 (1964).

Dunlop, D. J. Interactions in rocks and the reliability of paleointensity data, *Earth Planet Sci. Letters,* **7**, 178-182 (1969).

Dunlop, D. J. and G. F. West. An experimental evaluation of single domain theories, *Revs. Geophys.*, **7**, 709-757 (1969).

Néel, L. Some theoretical aspects of rock-magnetism, *Adv. Phys.*, **4**, 191-243 (1955).

Smit, J. and H. P. J. Wijn. "Ferrites," Wiley and Sons, New York, p. 367 (1959).

Stacey, F. D. The physical theory of rock magnetism, *Adv. Phys.*, **12**, 45-132 (1963).

Magnetic Minerals (Chapter 3)

Akimoto, S. and I. Kushiro. Natural occurrence of titanomaghemite and its relevance to the unstable magnetization of rocks, *J. Geomagn. Geoelect.*, **11**, 94-110 (1960).

Basta, E. Z. Some mineralogical relationships in the system $Fe_2O_3 - Fe_3O_4$ and the composition of titanomaghemite, *Econ. Geol.*, **54**, 698-719 (1959).

Brooks, P. J. and W. O'Reilly. Magnetic rotational hysteresis characteristics of red sandstones, *Earth Planet. Sci. Letters*, **9**, 71-76 (1970).

Chevallier, R., S. Mathieu and E. A. Vincent. Iron-titanium oxide minerals in layered gabbros of the Skaergaard intrusion, East Greenland II, *Geochim. Cosmochin. Acta*, **6**, 27-34 (1954).

Chevallier, R., J. Bolfa and S. Mathieu. Titanomagnétites et ilmenites ferromagnétiques, *Bull. Soc. Franc. Miner. Crist.*, **78**, 307-346 (1955).

Deer, W. A., R. A. Howie and J. Zussman. "Rock forming minerals," Vol. 5, Non-silicates, Longmans, London. p. 371 (1962).

Gorter E. W. Chemistry and magnetic properties of some ferrimagnetic oxides like those occurring in nature, *Adv. Phys.*, **6**, 336-361 (1957).

Hedley, I. G. Chemical remanent magnetization of the FeO OH, Fe_2O_3 system, *Phys. Earth Planet. Interiors*, **1**, 103-121 (1968).

Nicholls, G. D. The mineralogy of rock magnetism, *Adv. Phys.*, **4**, 113-190 (1955).

O'Reilly, W. Application of neutron diffraction and Mössbauer effect to rock magnetism, in "Application Modern Physics to Earth and Planetary Interiors" (Ed. S. K. Runcorn) Wiley-Interscience, p. 479-484 (1969).

Ramdohr, P. "The ore minerals and their intergrowths," Pergamon, Oxford, p. 1174 (1969).

Strangway, D. W., R. M. Honea, B. E. McMahon and E. E. Larson. The magnetic properties of naturally occurring goethite, *Geophys. J.*, **15**, 345-359 (1968).

Vincent, E. A. and R. Phillips. Iron-titanium oxide minerals in layered gabbros of the Skaergaard intrusion, East Greenland. I, *Geochim. Cosmochim. Acta*, **6**, 1-26 (1954).

Magnetism in Rocks (Chapter 4)

Bannerjee, S. K. and W. O'Reilly. The behaviour of ferrous ions in iron-titanium spinels, *J. Phys. Chem. Solids*, **28**, 1323-1335 (1967).

Briden, J. C. Ancient secondary magnetizations in rocks, *J. Geophys. Res.*, **70**, 5205-5221 (1965).

Collinson, D. W. Investigations into the stable remanent magnetization of sediments, *Geophys. J.*, **18**, 211-222 (1969).

Cox, A. Anomalous remanent magnetization of a basalt, *U.S. Geol. Surv., Bull.*, **1083**-E, 131-160 (1961).

Creer, K. M. Superparamagnetism in red sandstones, *Geophys. J.*, **5**, 16-28 (1961).

Creer, K. M. and N. Petersen. Thermochemical magnetization in basalts, *Z. Geophysik*, **35**, 501-516 (1969).

Doell, R. R. and P. J. Smith. On the use of magnetic cleaning in palaeointensity studies, *J. Geomagn. Geoelect.*, **21**, 579-594 (1969).

Dunlop, D. J., M. Ozima and H. Kinoshita. Piezomagnetization of single-domain grains: a graphical approach, *J. Geomagn. Geoelect*, **21**, 513-518 (1969).

Evans, M. E. and M. W. McElhinny. An investigation of the origin of stable remanence in magnetite-bearing igneous rocks, *J. Geomagn. Geoelect*, **21**, 757-773 (1969).

Graham, K. W. T. The re-magnetization of a surface outcrop by lightning currents, *Geophys. J.*, **6**, 85-102 (1961).

Griffiths, D. H., R. F. King, A. I. Rees and A. E. Wright. The remanent magnetization of some recent varved sediments, *Proc. Roy. Soc. Lond. A*, **256**, 359-383 (1960).

Irving, E. and A. Major. Post-depositional detrital remanent magnetization in an synthetic sediment, *Sedimentology*, **3**, 135-143 (1964).

Larson, E. E., M. Ozima, M. Ozima, T. Nagata and D. W. Strangway. Stability of the remanent magnetization of rocks, *Geophys. J.*, **17**, 263-292 (1969).

Malan, D. J. "Physics of lightning," English University Press, London, p. 176 (1963).

Néel, L. Théorie du trainage magnétique des ferromagnétiques au grains fins avec applications aux terres cuites, *Ann. Géophys.*, **7**, 90-102 (1949).

Petrova, G. N. Various laboratory methods of determining the geomagnetic stability of rocks, *Akad. Nauk. SSSR Izv. Geophys. Ser.*, 1585-1598 (1961).

Rees, A. I. The effect of water currents in the magnetic remanence and anisotropy of susceptibility of some sediments, *Geophys. J.*, **5**, 235-251 (1961).

Rimbert, F. Sur l'aimantation rémanente anhystéretique des ferromagnétiques, *C. R. Acad. Sci., Paris*, **245**, 406-408 (1957).

Rimbert, F. Contribution a l'étude de l'action de champs alternatifs sur les aimantations rémanentes des roches, *Rev. Inst. Franc. Petrol. Ann. Combust. Liquides*, **14**, 1 and 2 (1959).

Shimada, M., S. Kume and M. Koizumi. Demagnetization of unstable remanent magnetization by the application of pressure, *Geophys. J.*, **16**, 369-373 (1968).

Shive, P. N. Deformation and remanence in magnetite, *Earth Planet, Sci. Letters*, **7**, 451-455 (1970).

Verhoogen, J. The origin of thermoremanent magnetization, *J. Geophys. Res.*, **64**, 2441-2449 (1959).

Sampling and Measurement (Chapter 5)

As, J. A. Instruments and measuring methods in palaeomagnetic research, *Kon. Ned. Akad. Met. Inst., Med. Verh.,* **78**, p. 56 (1960).

Blackett, P. M. S. A negative experiment relating to magnetism and the Earth's rotation, *Phil. Trans. Roy. Soc. Lond. A,* **245**, 309-370 (1952).

Christie, K. W. and D. T. A. Symons, Apparatus for measuring magnetic susceptibility and its anisotropy, *Geol. Surv. Canada,* Paper 69-41, p. 10 (1969).

de Sa, A. and L. Molyneux. A spinner magnetometer, *J. Sci. Instr.,* **40**, 162-165 (1963).

Graham, J. W. Elimination of static charges from spinner magnetometers, *J. Geophys. Res.,* **73**, 4788 (1968).

Larochelle, A. The design of a spinner-type remanent magnetometer, *Geol. Surv. Canada,* Paper 64-43, p. 25 (1965).

Larochelle, A. and R. F. Black. The design and testing of an alternating-field demagnetizing apparatus, *Canad. J. Earth Sci.,* **2**, 684-695 (1965).

Statistics (Chapter 6)

Cox. A. Confidence limits for the precision parameter k, *Geophys. J.,* **18**, 545-549 (1969).

Doell, R. R. and A. Cox. Paleomagnetism of Hawaiian lava flows, *J. Geophys. Res.,* **70**, 3377-3405 (1965).

Fisher, R. A. Dispersion on a sphere, *Proc. Roy. Soc. Lond. A,* **217**, 295-305 (1953).

Irving, E., L. Molyneux and S. K. Runcorn. The analysis of remanent intensities and susceptibilities of rocks, *Geophys. J.,* **10**, 451-464 (1966).

Larochelle, A. A re-examination of certain statistical methods in palaeomagnetism, *Geol. Surv. Canada,* Paper 67-18, p. 15 (1967).

Larochelle, A. Further considerations on certain statistical methods in palaeomagnetism, *Geol. Surv. Canada,* Paper 67-26, p. 10 (1967).

Tarling, D. H. The magnetic intensity and susceptibility distribution in some Cenozoic and Jurassic basalts, *Geophys. J.,* **11**, 423-432 (1966).

Tarling, D. H. and D. T. A. Symons. A stability index of remanence in palaeomagnetism, *Geophys. J.,* **12**, 443-448 (1967).

Vincenz, S. A. and J. McG. Bruckshaw. Note on the probability distribution of a small number of vectors, *Proc. Camb. Phil. Soc,* **56**, 21-26 (1960).

Watson, G. S. A test for randomness of directions, *Mon. Not. Roy. Astr. Soc., Geophys. Suppl.,* **7**, 160-161 (1956).

Watson G. S. Analysis of dispersion on a sphere, *Mon. Not. Roy. Astr. Soc., Geophys. Suppl.,* **7**, 153-159 (1956).

Watson, G. S. and E. Irving. Statistical methods in rock magnetism, *Mon. Not. Roy. Astr. Soc., Geophys. Suppl.,* **7**, 289-300 (1957).

Wilson, R. L., S. E. Haggerty and N. D. Watkins. Variation of palaeomagnetic stability and other parameters in a vertical traverse of a single Icelandic lava, *Geophys. J.,* **16**, 79-96 (1968).

Geomagnetism (Chapter 7)

Bullard, E. C. and H. Gellman. Homogeneous dynamoes and terrestrial magnetism, *Phil. Trans. Roy. Soc. Lond. A.* **243**, 67-159 (1950).

Coe, R. Paleointensities of the Earth's magnetic field determined from Tertiary and Quaternary rocks, *J. Geophys. Res.,* **72**, 3247-3262 (1967).

Cox, A. Latitude dependence of the angular dispersion of the geomagnetic field, *Geophys. J.,* **20**, 253-269 (1970).

Cox, A. and R. R. Doell. Long period variations of the geomagnetic field, *Bull. Seism. Soc. Amer.,* **54**, 2243-2270 (1964).

Doell, R. R. History of the geomagnetic field, *J. Appl. Phys.,* **40**, 945-954 (1969).

DuBois, R. L. and N. Watanabe. Preliminary results of investigations made to study the use of Indian pottery to determine the paleointensity of the geomagnetic field for the United States AD600-1400, *J. Geomagn. Geoelect.,* **17**, 417-423 (1965).

Granar, L. Magnetic measurements on Swedish varved sediments, *Ark. f. Geofysik,* **3**, 1-40 (1958).

Hindmarsh, W. R., F. J. Lowes, P. H. Roberts and S. K. Runcorn (Eds.) "Magnetism and the Cosmos," Elsevier, New York, p. 436 (1967).

Jacobs, J. A. "The Earth's core and geomagnetism," Pergamon, Oxford, p. 137 (1963).

Opdyke, N. D. and K. W. Henry. A test of the dipole hypothesis, *Earth Planet. Sci. Letters,* **6**, 139-151 (1969).

Schwarz, E. J. A discussion of thermal and alternating field demagnetization methods in the estimation of palaeomagnetic field intensities, *J. Geomagn. Geoelect.,* **21**, 669-677 (1969).

Schwarz, E. J. and D. T. A. Symons. On the intensity of the paleomagnetic field between 100 million and 2500 million years ago, *Phys. Earth Plant. Interiors,* **1**, 122 128 (1968).

Smith, P. J. The intensity of the ancient geomagnetic field, a review and analysis, *Geophys. J.,* **12**, 321-362 (1967).

Thellier, E. and O. Thellier. Sur l'intensité du champ magnétique terrestre dans le passé historique et geologique, *Ann. Géophys.,* **15**, 285-376 (1959).

Rikitake, T. "Electromagnetism and the Earth's interior," Elsevier, Amsterdam, p. 308 (1966).

Watanabe, N. The direction of remanent magnetism of baked earth and its application to chronology for anthropology and archaeology in Japan, *J. Fac. Sci., Univ. Tokyo,* **2**, 1-188 (1959).

Watkins, N. D. A palaeomagnetic observation of Miocene geomagnetic secular variation in Oregon, *Nature,* **206**, 879-882 (1965).

Wilson, R. L. Permanent aspects of the Earth's non-dipole magnetic field over Upper Tertiary times, *Geophys. J.,* **19**, 417-437 (1970).

Reversals of Magnetisation (Chapter 8)

Ade-Hall, J. M. A correlation between remanent magnetism and petrological and chemical properties of Tertiary basalt lavas from Mull, Scotland, *Geophys. J.,* **8**, 404-423 (1964).

Ade-Hall, J. M. and N. D. Watkins. Absence of correlations between opaque petrology and natural remanence polarity in Canary Island lavas, *Geophys. J.,* **19**, 351-360 (1970).

Bhimasankaram, V. L. S. Partial magnetic self-reversal of pyrrhotite, *Nature,* **202**, 478-480 (1964).

Bullard, E. C. Reversals of the Earth's magnetic field, *Phil, Trans Roy. Soc. Lond. A,* **263**, 481-524 (1968).

Carmichael, C. M. The magnetic properties of ilmenite-haematite crystals, *Proc. Roy. Soc. Lond. A,* **263**, 508-530 (1961).

Cox, A. Lengths of geomagnetic polarity intervals, *J. Geophys. Res.,* **73**, 3247-3260 (1968).

Creer, K. M. and Y. Ispir. An interpretation of the behaviour of the geomagnetic field during polarity transitions, *Phys. Earth Planet. Interiors,* **2**, 283-293 (1970).

Heirtzler, J. R., X. Le Pichon and J. G. Brown. Magnetic anomalies over the Reykjanes Ridge, *Deep Sea Res.,* **13**, 427-443 (1966).

Helsley, C. E. and M. B. Steiner. Evidence for long intervals of normal polarity during the Cretaceous period, *Earth Planet. Sci. Letters,* **5**, 325-332 (1969).

Ishikawa, Y. and Y. Syono. Order-disorder transformation and reverse thermoremanent magnetism in the $FeTiO_3$-Fe_2O_3 system, *J. Phys. Chem. Solids,* **24**, 517-528 (1963).

McMahon, B. E. and D. W. Strangway. Investigation of the Kiaman magnetic division in Colorado redbeds, *Geophys. J.,* **15**, 265-285 (1968).

Nagata, T., S. Uyeda and S. Akimoto. Self-reversal of thermo-remanent magnetism in igneous rocks, *J. Geomagn. Geolelect.,* **4**, 22-38 (1952).

Néel, L. L'inversion de l'aimantation permanente des roches, *Ann. Géophys.,* **7**, 90-102 (1951).

Ozima, M. and E. E. Larson. Study of self-reversal of TRM in some submarine basalts, *J. Geomagn. Geoelect.,* **20**, 337-351 (1968).

Schult, A. Self-reversal of magnetization and chemical composition of titanomagnetites in basalts, *Earth Planet. Sci. Letters,* **4**, 57-63 (1968).

Verhoogen, J. Ionic ordering and self-reversal of magnetization in impure magnetites, *J. Geophys. Res.,* **61**, 201-209 (1956).

Watkins, B. D. and S. E. Haggerty. Primary oxidation variation and petrogenesis in a single lava, *Contr. Mineral. Petrol.,* **15**, 251-271 (1967).

Wilson, R. L. The paleomagnetism of baked contact rocks and reversals of the Earth's magnetic field, *Geophys. J.*, **7**, 194-202 (1962).

Wilson, R. L. Further correlations between the petrology and the natural magnetic polarity of basalts, *Geophys. J.*, **10**, 413-420 (1966).

Geology-Geophysics (Chapter 9)

Abdullah, M. I. and M. P. Atherton. The thermometric significance of magnetite in low grade metamorphic rocks, *Amer. J. Sci.*, **262**, 904-917 (1964).

Chamalaun, F. H. and K. M. Creer. Thermal demagnetization studies in the Old Red Sandstone of the Anglo-Welsh Cuvette, *J. Geophys. Res.*, **67**, 1607-1616 (1964).

Creer, K. M. Palaeozoic paleomagnetism, *Nature*, **219**, 246-250 (1968).

Creer, K. M. Review of palaeomagnetism, *Earth-Sci. Rev.*, **6**, 369-466 (1970).

Deutsch, E. R. Polar wandering and continental drift: an evaluation of recent evidence, in "Polar Wandering and Continental Drift," Amer. Assoc. Petrol Geol., 4-46 (1963).

Fuller, M. D. On the magnetic fabrics of certain rocks, *J. Geol.*, **72**, 368-376 (1964).

Glass, B., D. B. Ericson, B. C. Heezen, N. D. Opdyke and J. A. Glass. Geomagnetic reversals and Pleistocene chronology, *Nature*, **216**, 437-442 (1967).

Heirtzler, J. R, G. O. Dickson, E. M. Herron, W. C. Pitman III, and X. Le Pichon. Marine magnetic anomalies, geomagnetic field reversals, and motions of the ocean floor and continents, *J. Geophys. Res.*, **73**, 2119-2136 (1968).

Hospers, J. and S. I. van Andel. Paleomagnetism and tectonics, a review, *Earth-Sci. Revs.*, **5**, 5-44 (1969).

Kawai, N., K. Hirooka and T. Nakajima. Palaeomagnetic and potassium-argon age informations supporting Cretaceous-Tertiary hypothetic bend of the main island Japan, *Palaeogeog., Palaeoclimatol., Palaeoecol.*, **6**, 277-282 (1969).

Khan, M. A. The anisotropy of magnetic susceptibility of some igneous and metamorphic rocks, *J. Geophys. Res.*, **67**, 2873-2885 (1962).

King, R. F. Magnetic fabric of some Irish granites, *Geol. J.*, **5**, 43-66 (1966).

Mackenzie, D. P. Speculations on the consequences and causes of plate motions, *Geophys. J.*, **18**, 1-32 (1969).

McDougall, I. and R. F. Black. The use of magnetic measurements for the study of the structure of talus slopes, *Geol. Mag.*, **95**, 252-260 (1958).

Morgan, W. J. Rises, trenches and crustal blocks, *J. Geophys. Res.*, **73**, 1959-1982 (1968).

Nairn, A. E. M. (Ed.) "Descriptive palaeoclimatology," Interscience, New York, p. 380 (1961).

Nairn, A. E. M. (Ed.) "Problems in Palaeoclimatology," Interscience, New York, p. 705 (1964).

Picard, M. D. Paleomagnetic correlation of units within the Chugwater (Triassic) Formation west-central Wyoming, *Bull. Amer. Ass. Petrol. Geol.*, **48**, 269-291 (1964).

Rees, A. I. The use of anisotropy of magnetic susceptibility in the estimation of sedimentary fabric, *Sedimentology,* **4**, 257-271 (1965).

Runcorn, S. K. Paleomagnetic comparisons between Europe and North America, *Proc. Canad. Geol. Ass.,* **8**, 77-85 (1956).

Sasajima, S., J. Nishada and M. Shimada. Paleomagnetic evidence of a drift of the Japanese main island during the Paleogene period, *Earth Planet. Sci. Letters,* **5**, 135-141 (1968).

Symons, D. T. A. Paleomagnetic evidence on the genesis of the hard hematite ore deposits of Vermilion Range, Minnesota, *Canad. J. Earth Sci.,* **4**, 449-460 (1967).

Symons, D. T. A. The magnetic and petrologic properties of a basalt column, *Geophys. J.,* **12**, 473-490 (1967).

Tarling, D. H. The palaeomagnetic evidence of displacements within continents, in "Time and Place in Orogeny" (Eds. Kent, P. E., G. E. Satterthwaite and A. M. Spencer) Geol. Soc. London, 95-113 (1969).

Tarling, D. H. Palaeomagnetism and the origin of the Red Sea and Gulf of Aden, *Phil. Trans. Roy. Soc., Lond., A.* **276**, 219-226 (1970).

Van Andel, S. I. and J. Hospers. Paleomagnetism and the hypothesis of an expanding Earth: a new calculation method and its results, *Tectonophysics,* **5**, 273-285 (1968).

Van der Voo, R. and J. D. A. Zijderveld. Paleomagnetism in the western Mediterranean area, *Verh. Kon. Ned. Geol. Mijn., Gen.* **26**, 121-138 (1969).

Index

The main page references, e.g. definition and detailed description, are in bold figures where several pages are indexed. The examples of various applications are not indexed, i.e. no index is given for the Alps, Japanese island arc, etc., as this book is not intended as a summary of previous palaeomagnetic results.

Acid treatment 46, 47
age of remanence 32, 41, 42, **48-51**, 53, 92, 124
alidade 59
alternating magnetic field 17, 43-45, 66, **67-69**, 87, 104
amorphous Fe-Ti oxides & hydroxides, 26, 27
amphiboles 26
angle of rest 60
angular standard deviation 79
anhysteretic magnetisation **17**, 40, 43, 64, 68, 69, 85
anisotropy **18-20**, 32, 37, 38, 43, **52-53**, 62, 64, **65, 66**, 147, 151
antiferromagnetism 8, 9, 23, 26-28, 115
archaeology 60, 98-101, 104, 106-108
astatic magnetometer 61-62, 64
atomic magnetisation 5-9
attempted polarity reversals 123, 128
axial geocentric dipole 38, 41, 43, 96, 97, **107-112**, 123, 129, 137
baked rocks and soils 3, 37, 124
bedding correction — see tilt correction
Bemmelen, W. van 2
between-site scatter **84**, 102, 103
Bloch wall **9, 10**, 11, 12, 14, 18
blocking temperature **15**, 16, 17, 33, 41
blocking volume **15**, 17, 33
Bohr magneton 5
Broun, J. A. 3

Brunhes, B. 3
burial, effect of 18, 27, 36, 37, 51, 54, 60, 92, 151
cassiterite 29
Castro, J. de 2
cement in sediments 27, **36**, 46, 47
chemical magnetisation **17**, 24, 27, 28, 33, 35-37, 43, 51, 85, 102
Chevallier, R. 4
chromite 28
circular standard deviation **77, 78,** 79, 103, 130
circular standard error 78, 79
clay minerals 27, 29, 36
climate, see palaeoclimates
closure domains **9**, 11, 12, 14
coal 148, 149
cobalt 7
coercivity 12, 14, 29, 31, 40, 41, 45, 46, 69. 86, 87
colatitude 74
columnar jointing 34
compaction, see burial
compass 1-3, 40, 59
confidence, cone of **77**, 78, 79, 130
    ellipse of **81**, 82, 83
conglomerate stability test 48, 49
consistency stability test 41
contact test, see baked rocks
continental drift, see also plates 143
convection 112, 140, 141
core of the Earth 1, 112, 113, 123, 141

cosmic radiation 99
crust 136
crystal lattice 7, 8, **9**, 10, 12, 14, 18-20, 22, 23, 46, 51, 115-117
crystalline anisotropy **18**, 19, 26, 52, 53
cubanite 28
Curie temperature **7**, 9, 15, 20, 23-25, 28-30, 33, 34, 57, 112, 116, 150
dating rocks with palaeomagnetism 114, **129-136**
David, P. 3
declination 72
deep-sea sediments 36, 55, 108, 109, 119, 121, 124, 127, 151
Delesse, A. 3
demagnetisation factor 19
demagnetisation of remanence 17, 40, **43-45**, 52, **66-69**, 71, 85, 86
density fractionation 29
depositional magnetisation 33, **35**, 107
detrital particles 27, 28, 35, 36, 46, 47
diagenesis **27**, 33, 36, 38, 92
diamagnetism **6**, 7
dip 58
dip poles 93
dipole nature of geomagnetic field 2, 3, 96, 97, 102, **107-112**, 129, 137
direct exchange interaction 7
directions, calculation of 72, **73**, 81, 83
domains **9**-14, 33, 34, 38, 40, 41, 43, 45, 46
domain wall, see Bloch wall
drilling 56-60
dynamo, geomagnetic 112, 113
'easy' axes of magnetisation **9**, 13-15, 18
electron, magnetisation of 5
energy barriers in crystals **10**, 12, 14, 18
epochs, polarity 119, 120
events, polarity **119, 120**, 121, 151
evolution and reversals 127, 128
exchange interactions 7
expanding Earth hypothesis 141-143
exsolution **23-26**, 33, 35, 38, 41, 51, 52, 117, 118

extraterrestrial rocks 9, 21, 29, 60, 93, 113
ferrimagnetism 8, 9, 23, 25, 28, 115, 117
ferromagnetism (s.s.) **8**, 28
ferromagnetism (s.l.) **7**, 15, 32, 33
field-free space 67, **69, 70**
Fisher, R. A. 4, 75
Fisherian distribution **75-81**, 90, 91
flow banding 52, 147
fluxgates 64, 66, 70
folds and fold tests, see also tilt correction 48, 49, 60
Folgerhaiter, G. 3
foliation 147
Forstermann, M. 3
fossils, distribution of 148, 149
    extinction of 127, 128
franklinite 28
'free' poles **9**, 13
furnaces for demagnetisation 67
Gauss, K. F. 3, 98
Gaussian distribution 75, 77, 78, 87
Gellibrand, H 2
geomagnetic field, present 93-98
Gilbert, W. 2
glaciation 148
goethite **27**, 30, 40, 116
gyrocompass 59
haematite 9, 18, 22, 25, **26**, 27, 29, 31, 36, 45-47, 51, 67, 69, 87, 99, 107, 116, 117, 147
Halley, E. 2
Hartmann, G. 2
Haruna dacite 117
helmholtz coils 70
homogeneity, see inhomogeneity
hydrostatic pressure **18**, 20, 37
hydroxides of iron 27, 36, 40
hysteresis loop **12, 13**, 14
igneous rocks 3, 4, 17, **21-27**, 29, 32, **33-35**, 40, 41, 45, 50-53, 55, 59, 69, 87-89, 92, 98, 99, 101-107, 109, 117-119, 123, 150, 151
ilmenite 22, 24, **25**, 26, 117, 118

ilmeno-haematite series 22, **24**, **25**, 26,
    116, 117, 150
ilmeno-rutile 22
imperfections 9, 12, 24
impurities 12, 24, 25, 31, 116
inclination 72
inclination error **35**, **36**, 92
inclined geomagnetic dipole 93, 95, 102
induced anisotropy 20
inhomogeneity 33, 53, 57, 62-64, **66**, 85,
    90
instrumental error 43, 57, 70, 71, 84, 85,
    102, 104
intensity of past magnetic fields 50, 51,
    89-90, 92, **104-107**, 123
intensity, statistical analysis of 89
intergrowths 21, 24, 25
intersection of meridians 92, **110-111**
ionospheric currents 93, 128
iron 7, 9, 21, 29, 31, 113
iron ores, origin of 135
island arcs 143, 145
isothermal remanence **12**, 39, 40,
    45, 46
isotropy, see anisotropy
jacobsite 28
Kiaman reversed epoch 120
lightning 3, 4, 34, **38-40**, 59, 89
limonite **27**, 40
lithosphere **136**, 140, 141
lodestone 1-3
log-normal distribution 87-89
low temperature effects 18, 46, 47
lunar samples 9, 113
maghemite 22, **24**, 26, 151
magnesioferrite 28
magnetic separation 29
magnetic shielding 69, 70
magnetic storms 34, 93
magnetite 9, 16, 17, **21-24**, 29, 30, 46, 47,
    116, 118, 147, 150
magnetocrystalline anisotropy 52
magnetocrystalline forces **9**, 10, 14
magnetometers **60-64**, 66, 67

magnetostatic forces **9**, 10, 11, 14, 19,
    115, 116
mantle 113, 119, 136, 140, 141, 151
Matuyama, M. 4
mechanical remanence 17
Melloni, M. 3
Mercator, G. 2
Mercanton, P. 4
metamorphic rocks 18, 20, **26-27**, **37-38**,
    41, 52, 53, 150
meteorites 9, 21, 113, 128
mica 29, 52
microscope & microprobe 30, 31
moon, see lunar samples
Morin transition **26**, 46
mountain systems & palaeomagnetism
    143-146, 151
mossbauer techniques 29
multidomain, identification 31, 35
nappe structures 143, 146
natural remanence **32**, 41, 45, 46, 50, 55,
    104, 105, 107
Neckham, A. 2
Néel, L. 4
Néel temperature 25
nickel 7, 9, 21, 29, 113
non-dipole field **97**, 98, 102, 123
normal distribution 75
Norman, R. 2
oil 148-150
olivine 26, 35, 36
orbital magnetisation 5, 6
ore deposits 135, 146
orientation 56, **57-60**, 70, 84, 85, 102
oroclines 143-146
oxidation 24, 26-29, 36, 40, 67, 69,
    116-119, 151
palaeoclimates 112, 128, 148, 149
palaeointensity, see intensity
palaeolatitudes **74**, 89, 103, 107, 111, 112,
    136-138, **147-150**
paramagnetism **6**, 7, 9, 15, 26, 28, 32
parasitic ferromagnetism **9**, 25-27, 117
partial thermoremanence 34, 104

Peregrinus, P. 2
petrofabric analysis 147
petrology : polarity correlation 114,
    117-119
pilot specimens **45**, 85
plate tectonics and motions **136**, 138, 140,
    141
polarity changes, see reversals
polarity time scale 114, **119-121**, 122,
    123-127, 131, 132
polar wander, true 139-141
polar wandering curves **132**, 133, 135,
    137-140
pole position, calculation of 72, 73, **74**,
    81, 83, **110-111**
poloidal field 112-113
precision 75, 77-83, 130
precision parameter **75, 77**, 79, 83, 86
pressure remanence, see also uniaxial &
    hydrostatic pressure 18
primary magnetisation 32-37, 41, 51, 53,
    55, 85, 89
projections 74
pseudobrookite 22, 26
pyrite 28
pyroxene 26, 35, 36
pyrrhotite **28**, 116, 147
radio-active dating 51, 57, 99, 119, 120
radius of the Earth 141-143
randomness, tests of 75
rare earths 28, 29
reconstructions, continental 136-139
red sediments 28, 36, 67, 69, 149
relaxation time **14**, 15-17, 33, 34, 37, 40,
    41, 43, 45, 48, 52, 66, 67, 69, 71
remagnetisation, circles of 42, 43
remanent magnetisation, see chemical;
    depositional; isothermal; natural;
    thermo-; viscous.
reversals of polarity, general 3, 4, 38, 41,
    112, 113, **114-128**, 129, 131, 132,
    136
rifts and ridges, oceanic 124, 125, 127,
    136, 140, 141, 150, 151

rotations, tectonic 48, 60, 111, **136-147**
rutile 22, 26
samples **56**, 57-60, 70, 71, 84-86, 90
saturation magnetisation **12**, 13, 24, 25,
    29, 40, 46
scatter, measures of 75-79
scree slopes 146
sea-floor spreading **124-127**, 136, 140,
    141, 151
secondary magnetisation **32**, 35, 37,
    38-40, 43, 46, 51, 85, 86, 107, 143
secular variation 2, 28, 92, 97, **98-104**,
    106, 107, 111-113, 129, **130**, 141
sedimentary rocks 21, **27**, **28**, 29, 32, 33,
    35-37, 41, 45-47, 51-53, 55, 56, 60,
    67, 69, 85, 87-89, 92, 100, 102,
    107, 109, 123, 127
self-reversals 20, 25, 47, **114-119**, 123,
    124
shape anisotropy **19**, 20, 52, 53
significance points **75, 76**, 78
single domains **13**, 14, 31, 35
site **55**, 56, 60, 71, 84-86, 90, 92, 102,
    103
slump test of stability 48, 49
smeared distributions 41-43
solar radiation 92, 127, 128
solid solutions, see ilmenohematite and
    titanomagnetite series
specimen **56**, 57, 61, 64, 68-70, 84-86
spherical harmonic analysis 3, 96-98
spin magnetisation 5, 6, 8, 13
spinner magnetometer 61, **62-64**, 66
spontaneous magnetisation **7**, 8, 9, 15
stability, magnetic 3, 14, **40-47**, 48, 52,
    53, 71, 85, **86, 87**, 90, 115
    factor 86-87
    index **87**, 90
standard deviation 89
standstills, during reversals 123, 132
statistical analysis 56, **72-91**, 130
steady field demagnetisation 45
step demagnetisation, see demagnetisation
storage 38, 41

strung distribution, see smeared
    distribution
strike **58**, 60
submarine igneous rocks 33, 117, 124-127,
    132, 136, 151
sulphide minerals 27, 28
sun compass 59
super exchange interactions **7**, 8, 116
superparamagnetism **15**, 17
susceptibility **6**, 7, 12, 30, 34, 66, 67,
    87-89
susceptibility bridge 64, 65
temperature, effects of changes 6, 7,
    14-17, 18, 19, 34, 35, 37, 40, 151
terella 2
thermal demagnetisation of rocks 43-45,
    66, 67, 87, 104, 105
thermoremanence 3, **16**, 33-35, 37, 38,
    43, 50, 51, 67, 99
thrusts 146
tilt correction and test 42, 48, 60, 71, 84,
    92
time, effect on remanence 10, **14-17**
titanomagnetite series **22-24**, 26, 31, 36,
    51
topographic orientation 59
toroidal field 112, 113

torque magnetometer 65
torsion fibre 61, 62
transient variations 93
transition polarity zones 107, **121-123**,
    124, 127, 129, 131, **132**
transition temperatures **18**, 46,
    47, 116
trenches, oceanic 136, 140
trevorite 28
troilite 28
tumblers 68, 69
ulvite = ulvospinel 22, 24
uniaxial pressure 18, 20, 37, 52-54, 151
varves, glacial 36, 100, 101
viscous remanence **14**, 29, 34, 35, 37, **38**,
    40, 43, 50, 51, 53, 89, 104, 113,
    136, 151
volume, effect of changes **14**-17
within-site precision 84, 102
weathering 21, 26, 30, 38, **40**, 57, 117,
    151
weighting systems in statistics 73, 85, 90,
    91
westward drift 98, 102, 113
Wilcke, J. C. 2
wustite 22
X-ray diffraction & fluorescence 29